电化学生物传感器在细胞活性氧和活性氮检测中的应用研究

高丽霞 王小登 著

北京
冶金工业出版社
2024

内 容 提 要

本书围绕近年来电化学生物传感器用于评价抗肿瘤药物的生物活性及作用机制进行叙述，重点介绍了电化学生物传感器的发展现状及趋势，活性氧的生物学功能与作用，超氧阴离子电化学传感器、一氧化氮电化学传感器和过氧化氢电化学传感器的制备与性能研究，以及当前最新的活性氧电化学传感器在细胞生物学中的应用。

本书可供化学、生物学专业的科研人员、高校教师和研究生阅读参考。

图书在版编目（CIP）数据

电化学生物传感器在细胞活性氧和活性氮检测中的应用研究／高丽霞，王小登著． -- 北京：冶金工业出版社，2024.9． -- ISBN 978-7-5024-9976-1

Ⅰ．Q2

中国国家版本馆 CIP 数据核字第 20249ZB280 号

电化学生物传感器在细胞活性氧和活性氮检测中的应用研究

出版发行	冶金工业出版社	电 话	(010)64027926
地 址	北京市东城区嵩祝院北巷39号	邮 编	100009
网 址	www.mip1953.com	电子信箱	service@mip1953.com

责任编辑　夏小雪　　美术编辑　吕欣童　　版式设计　郑小利
责任校对　李欣雨　　责任印制　禹　蕊
北京印刷集团有限责任公司印刷
2024年9月第1版，2024年9月第1次印刷
710mm×1000mm　1/16；11 印张；188 千字；165 页
定价 78.00 元

投稿电话　(010)64027932　　投稿信箱　tougao@cnmip.com.cn
营销中心电话　(010)64044283
冶金工业出版社天猫旗舰店　yjgycbs.tmall.com
（本书如有印装质量问题，本社营销中心负责退换）

前　言

活性氧和活性氮是生物体内一类非常不稳定的含氧化合物的总称，属于有氧代谢过程中形成的正常产物。正常情况下在生物体内细胞代谢需要活性氧，但是当机体氧化还原平衡失调时活性氧和活性氮产生增多，导致细胞损伤，引起机体病变。目前活性氧和活性氮的检测方法通常采用传统方法间接评价活性氧水平，这种方法具有耗时、检测费用高等不足。随着生物学的快速发展，生物电化学传感器作为化学传感器中研究最早、最深入的一个分支，具有检测特异性强、便捷、费用低等优点。电化学传感器用于活性氧和活性氮的分析检测对获取生命过程中的化学与生物信息、了解生物分子及其结构与功能的关系、阐述生命活动的机理以及对疾病的有效诊断与治疗都发挥着十分重要的作用。

本书共7章。其中，第1章概述了电化学生物传感器的研究现状；第2章介绍了纳米酶电化学生物传感器在活性氧和活性氮检测中的应用；第3章介绍了电化学生物传感器在药物开发中的应用；第4章重点介绍了电化学生物传感器在UV诱导黑色素瘤细胞O_2^{-}的应用；第5章重点介绍了电化学生物传感器在细胞NO检测中的应用研究；第6章阐述了CGMA/纸夹芯电化学传感器在细胞过氧化氢检测中的应用；第7章简述了电化学生物传感器的应用展望。

本书根据作者在生物电化学传感器与肿瘤细胞生物学研究领域的交叉与融合研究中积累的教学科研经验，并参考国内外该领域的众多图书资料及学术论文撰写而成。本书不仅具有较高的理论参考价值，而且具有广泛的应用价值，它既可为科研部门相关专业的科研人员提供学术参考，也可供高等院校相关专业的本科生或研究生作为教学用书或参考书使用。

　　本书的出版得到了国家自然科学基金、重庆市科学技术局、重庆市教育委员会与重庆文理学院等的共同资助。由于作者的学识水平所限，书中不妥之处，还望读者给予批评指正！

作　者

2024 年 5 月于重庆文理学院

目 录

1 绪论 ·· 1

　1.1 电化学生物传感器概述 ·· 1
　　1.1.1 电化学生物传感器的工作原理 ··· 1
　　1.1.2 电化学生物传感器的分类 ··· 2
　　1.1.3 电化学生物传感器的表征方法 ··· 8
　　1.1.4 电化学生物传感器的应用现状 ·· 10
　1.2 电化学生物传感器的发展趋势 ·· 11
　1.3 电化学活性氧和活性氮传感器概述 ·· 13
　1.4 小结 ·· 13
　参考文献 ··· 14

2 纳米酶电化学生物传感器在活性氧和活性氮检测中的应用 ································ 21

　2.1 引言 ·· 21
　2.2 细胞活性氧和活性氮的生物学功能 ·· 21
　　2.2.1 细胞中活性氧和活性氮（RONS）概述 ··· 22
　　2.2.2 细胞中活性氧和活性氮的检测 ··· 28
　　2.2.3 细胞中活性氧和活性氮的清除 ··· 32
　　2.2.4 细胞中活性氧和活性氮对机体的危害 ··· 35
　　2.2.5 细胞活性氧和活性氮在抗肿瘤药物研究中的应用 ···································· 37
　2.3 基于纳米酶的电化学生物传感器在活性氧检测中的应用 ······························· 38
　　2.3.1 基于过渡金属的纳米酶电化学生物传感器 ·· 39
　　2.3.2 基于贵金属的纳米酶电化学生物传感器 ·· 42
　　2.3.3 基于金属有机骨架的纳米酶电化学生物传感器 ······································· 43
　2.4 电化学生物传感器在活性氮检测中的应用 ··· 44

2.5 小结 ·· 45

参考文献 ·· 46

3 电化学生物传感器在药物开发中的应用 ·· 57

3.1 引言 ·· 57

3.2 生物学研究中的电化学参数与技术 ··· 59

3.3 电化学方法在体外评价药物代谢中的应用 ·· 60

3.4 电化学生物传感器在药物代谢中的研究 ··· 60

 3.4.1 酶标记的电化学传感器在药物代谢物检测中的应用 ··················· 61

 3.4.2 纳米酶标记的电化学传感器在药物研究中的应用 ······················ 62

3.5 电化学微流控技术在药物代谢中的应用 ··· 65

3.6 化学-质谱联用检测药物代谢物 ·· 66

3.7 电化学方法在抗肿瘤药物发现中的应用 ··· 67

3.8 电化学生物传感器在药物开发中的局限性与挑战 ·································· 72

3.9 小结 ·· 73

参考文献 ·· 73

4 电化学生物传感器在 UV 诱导黑色素瘤细胞 $O_2^{·-}$ 的应用 ·························· 85

4.1 引言 ·· 85

4.2 结果与讨论 ·· 87

 4.2.1 电化学检测装置 ·· 87

 4.2.2 CNT/DNA@$Mn_3(PO_4)_2$ 纳米复合材料制备 ································ 87

 4.2.3 超氧阴离子电化学传感器量化紫外光诱导黑色素瘤细胞产生的 $O_2^{·-}$ ·· 89

 4.2.4 HaCaT 和 A375 细胞在紫外光照射下的生长能力 ······················ 92

 4.2.5 生育酚可以减少紫外线诱导黑色素瘤细胞产生的 $O_2^{·-}$ ··············· 92

 4.2.6 电化学传感器预测生育酚在细胞成活能力中的保护作用 ············ 97

 4.2.7 对紫外线诱导细胞氧化损伤时不同检测方法比较 ······················ 99

4.3 小结 ··· 103

参考文献 ··· 104

5 电化学生物传感器在细胞 NO 检测中的应用研究 …… 109

5.1 引言 …… 109

5.2 rGO/CeO$_2$ 纳米复合材料电化学生物传感器的构建 …… 111

5.3 rGO-CeO$_2$ 电化学传感器检测肝癌细胞中 NO 的水平 …… 114

    5.3.1 基于 rGO-CeO$_2$ 的 NO 电化学传感器的校正 …… 114

    5.3.2 基于 rGO-CeO$_2$ 的 NO 电化学传感器的稳定性与选择性 …… 115

    5.3.3 索拉非尼在抑制肝癌细胞增殖过程中诱导细胞产生 NO …… 116

    5.3.4 FGF19 影响索拉非尼刺激肝癌细胞产生的 NO 水平 …… 118

    5.3.5 NO 电化学传感器检测肝癌细胞释放的 NO 水平 …… 122

    5.3.6 在索拉非尼耐药细胞中 NO 水平影响耐药细胞对索拉非尼的敏感性 …… 122

    5.3.7 电化学方法评价 BLU9931 增强索拉非尼对耐药细胞的抑制能力 …… 123

    5.3.8 rGO-CeO$_2$ 电化学生物传感器在肝癌治疗中的潜在应用价值 …… 126

5.4 金纳米颗粒-3D 石墨烯水凝胶纳米复合材料电化学传感器检测细胞释放 NO …… 128

    5.4.1 Au NPs-3DGH 复合材料的制备及其电极修饰 …… 129

    5.4.2 Au NPs-3DGH 纳米复合物的一氧化氮催化性能 …… 130

    5.4.3 Au NPs-3DGH 纳米复合物传感膜的选择性 …… 133

    5.4.4 Au NPs-3DGH 纳米复合物传感膜对活细胞的实时监测 …… 133

    5.4.5 Au NPs-3DGH 电化学传感器应用展望 …… 136

参考文献 …… 137

6 CGMA/纸夹芯电化学传感器在细胞过氧化氢检测中的应用 …… 143

6.1 引言 …… 143

6.2 过氧化氢电化学生物传感器 …… 144

    6.2.1 碳纳米管/石墨烯/二氧化锰气凝胶（CGMA）的制备 …… 144

    6.2.2 碳纳米管/石墨烯/二氧化锰气凝胶（CGMA）电极的表征 …… 145

    6.2.3 CGMA 电极/纸夹芯装置的制作 …… 149

    6.2.4 CPGA 电极/纸夹芯装置的电化学表征 …… 149

6.3 基于 CGMA 电极/纸夹芯装置的电化学传感器检测过氧化氢 …………… 150
6.4 基于 CNT/石墨烯/MnO_2 纳米复合物与纸夹芯装置的电化学传感器
    原位监测活细胞释放 H_2O_2 …………………………………………… 153
参考文献 ………………………………………………………………………… 155

7 电化学生物传感器的应用展望 ……………………………………………… 159
7.1 引言 …………………………………………………………………………… 159
7.2 新型电化学生物传感器 ……………………………………………………… 160
7.3 未来电化学生物传感器的发展趋势 ………………………………………… 161
7.4 电化学生物传感器面临的挑战 ……………………………………………… 162
7.5 小结 …………………………………………………………………………… 163
参考文献 ………………………………………………………………………… 164

1 绪 论

1.1 电化学生物传感器概述

电化学生物传感器[1]是一种利用电化学检测方法检测生物分子的小型器件，可以用于实时检测生物分子，属于一类特殊的传感器，它以生物活性单元（如酶、抗体、核酸、细胞等）作为生物敏感单元[2,3]，利用这些固定化的生物敏感物质与适当的电化学信号转换器组成的生物电化学分析系统，对生物分子具有特异的识别能力，并且能够检测生物分子与分析物之间的相互作用，将生物反应所引起的化学、物理变化转换成可检测的信号。它具有高灵敏度、低检测限、快速响应和恢复等优点。生物传感器是由生物、化学、物理、医学、电子技术等多种学科相互渗透与融合而发展起来的高新技术[4]，能够对所需要检测的物质进行快速分析和追踪[5]。生物传感器作为电化学传感器的重要分支之一，它是目前研究最深入、最完善的，而且它也是科学家的关注焦点和科学技术发展以及社会发展需求相互作用的结果。迄今为止，电化学生物传感器已经成为一个涉及内容广泛、多学科交叉、充满创新活力的新领域，一些研究成果已在生物技术、食品工业、临床检测、医药工业、生物医学、环境分析等领域获得实际应用[6-8]。在此，主要讨论生物小分子活性氧电化学传感器的制备及其在抗肿瘤药物研究中的应用。

1.1.1 电化学生物传感器的工作原理

电化学生物传感技术是指将生物物质与电化学传感器相结合，利用生物物质的生物化学反应过程对电化学传感器进行信号转化的技术，用于检测生物分子及其相互作用等生化过程的一种分析仪器。其工作原理基于生物体系内化学反应所产生的电子转移过程，使传感器产生相应的电化学信号，采用电化学信号进行检测和分析。通俗地说，其核心是将生物分子与电化学检测相结合，利用电化学反应的变化来检测生物分子的存在而产生信号。具体来说，电化学生物传感器的工作原理主要分为以下几个步骤。(1) 生物识别分析：选择适当的生物识别，如酶、抗体、DNA 等，与待测分子相互作用，产生一个特定的生物电化学反应。(2) 转化信号：生物分子的特异性与电化学传感器的变化相联系，产生相应的电化学信号（如电流、电势等），经过相应的检测方式转化成数字信号。(3) 分析测量：通过电化学测量、分析、计算等方式计算出待测物质的浓度或其他相关参数。总体上，电化学生物传感器的工作原理是将待测生物分子与特异的分析生物识别物相结合，转化为与电化学传感器相联系的电化学信号，从而实现对于待测物的定量分析。在实际应用中，该技术常用于生物医学检测、生态环境检测等领域。

电化学生物传感器的基本构成如图 1.1 所示。

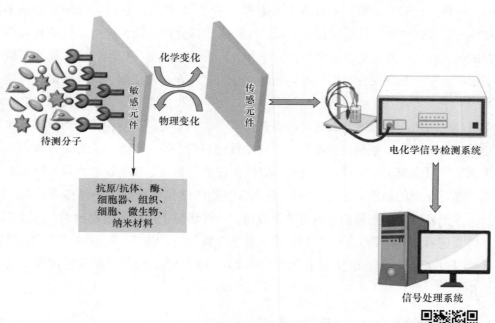

图 1.1　电化学生物传感器的基本构成

图 1.1 彩图

1.1.2　电化学生物传感器的分类

由于使用生物材料作为传感器的敏感元件，所以电化学生物传感器具有高度选择性，是快速、直接获取复杂体系组成信息的理想分析工具。

根据作为敏感元件所用生物材料的不同，电化学生物传感器可分为酶电极传感器、微生物电极传感器、电化学免疫传感器、组织电极与细胞器电极传感器、电化学 DNA 传感器、电化学非酶型纳米材料传感器等（见图 1.2），电化学非酶型纳米材料传感器已成为当今电化学生物传感器发展的趋势，在此我们将对电化学非酶型纳米材料传感器进行重点阐述。

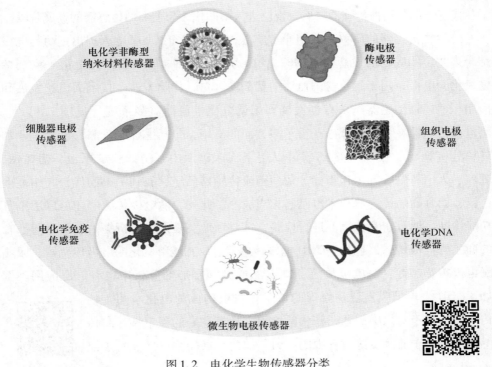

图 1.2　电化学生物传感器分类

图 1.2 彩图

1.1.2.1　酶电极传感器

酶电极传感器是一种特殊的化学传感器，它以生物活性单元（如酶）作为生物敏感基元，对被测目标物具有高度选择性。这种传感器通过捕捉目标物与敏感基元之间的反应所产生的可测信号，实现对目标物的定量测定。以葡萄糖氧化酶（GOD）电极为例简述其工作原理。在 GOD 的催化下，葡萄糖（$C_6H_{12}O_6$）被氧化生成葡萄糖酸（$C_6H_{12}O_7$）和过氧化氢。根据上述反应，可通过过氧化氢电极（检测氧的消耗）、过氧化氢电极（检测 H_2O_2 的产生）和 pH 电极（检测酸度变化）来间接测定葡萄糖的含量。因此，只要将 GOD 固定在上述电极表面即可构成检测葡萄糖的 GOD 传感器，这便是所谓的第一代酶电极传感器。这种传感器由于是间接测定法，故干扰因素较多。第二代酶电极传感器是采用氧化还原电子媒介体在酶的氧化还原活性中心与电极之间传递电子。因此，第二代酶电极传感器不受测定体系的限制，测量浓度线性范围较宽，干扰少。现在不少研究者又在努力发展第三代酶电极传感器，即酶的氧化还原活性中心直接和电极表面交换电子的酶电极传感器[9,10]。葡萄糖传感器在几十年的发展中取得了重大进展，经历了三代基于酶葡萄糖传感器之后，现已进入第四代无酶葡萄糖传感器的发展阶段。

酶生物传感器的研究和应用非常广泛，自1962年Clark等人提出把酶与电极结合来测定酶底物的设想后，1967年Updike和Hicks研制出世界上第一支葡萄糖氧化酶电极，用于定量检测血清中葡萄糖含量，伴随着这一技术的迅速发展和应用，葡萄糖传感器已从有创发展到无创电化学血糖监测系统（见图1.3）[11]。酶电极传感器在多个领域中获得了多方面的应用，包括医疗、食品、发酵工业和环境分析等[12]。酶电极传感器可以用于临床医学中，例如测定尿素、葡萄糖、乳酸、天门冬酰胺等生化指标。葡萄糖酶传感器已经发展到了第四代，应用范围广泛，而乳酸酶传感器技术也已经相当成熟。此外，酶电极传感器还可以用于肾功能诊断，通过尿素传感器进行检验，针对性地实施人工透析等治疗。目前已有的商品酶电极传感器包括：GOD电极传感器、L-乳酸单氧化酶电极传感器、尿酸酶电极传感器等。酶电极传感器在食品成分和添加剂分析中也有广泛应用。例如，GOD-壳聚糖膜复合物修饰的铂工作电极构成的电化学酶生物传感器，用于水果中的葡萄糖含量检测，具有良好的稳定性。酶传感器可以用于监测发酵过程中的参数，如葡萄糖的浓度。这对于控

图1.3 彩图

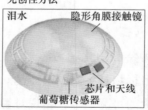
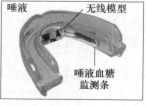

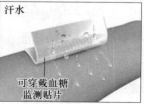

图1.3　葡萄糖传感器已从有创发展到无创电化学血糖监测系统

制发酵过程、提高产品质量和产量具有重要意义。酶电极传感器还可以用于水质监测，例如通过固定化多酚氧化酶研制成的多酚氧化酶传感器，可以快速测定出水中质量分数极低的酚类化合物。此外，还有研究利用酶抑制原理的电化学生物传感器在药物检测中的应用，展示了其在药物检测领域的潜力。综上所述，酶电极传感器因其具有高选择性和高灵敏度，在医疗、食品、发酵工业和环境分析等多个领域中发挥着重要作用。

1.1.2.2 微生物电极传感器

由于离析酶的价格昂贵且稳定性较差，限制了其在电化学生物传感器中的应用，从而使研究者想到直接利用活的微生物来作为分子识别元件的敏感材料。这种将微生物（常用的主要是细菌和酵母菌）作为敏感材料固定在电极表面构成的电化学生物传感器称为微生物电极传感器。其工作原理大致可分为三种类型：其一，利用微生物体内含有的酶（单一酶或复合酶）系来识别分子，这种类型与酶电极类似；其二，利用微生物对有机物的同化作用，通过检测其呼吸活性（摄氧量）的提高，即通过氧电极测量体系中氧的减少间接测定有机物的浓度；其三，通过测定电极敏感的代谢产物间接测定一些能被厌氧微生物所同化的有机物。

微生物电极传感器在发酵工业、食品检验、医疗卫生等领域都有应用[13,14]。例如：在食品发酵过程中测定葡萄糖的佛鲁奥森假单胞菌电极[15-17]，将固定化大肠杆菌215和氧电极组成微生物传感器用于测定维生素B12[18]。微生物电极传感器由于价廉、使用寿命长而具有很好的应用前景，然而它的选择性和长期稳定性等还有待进一步提高。

1.1.2.3 电化学免疫传感器

抗体对相应抗原具有唯一性识别和结合功能。电化学免疫传感器就是利用这种识别和结合功能将抗体或抗原和电极组合而成的检测装置[19]。根据电化学免疫传感器的结构可将其分为直接型和间接型两类。直接型的特点是在抗体与其相应抗原识别结合的同时将其免疫反应的信息直接转变成电信号。这类传感器在结构上可进一步分为结合型和分离型两种。前者是将抗体或抗原直接固定在电极表面上，传感器与相应的抗体或抗原发生结合的同时产生电势改变；后者是用抗体或抗原制作抗体膜或抗原膜，当其与相应的配基反应时，膜电势发生变化，测定膜电势的电极与膜是分开的。间接型的特点是将抗原和抗体结合的信息转变成另一种中间信息，然后再把这个中间信息转变成电信号。这类传感器在结构上也可进一步分为两种类型：结合型和分离型。前者是将抗体或抗原固定在电极上，而后者抗体或抗原和电极是完全分开的。间接型电化学免疫传感器通常是采用酶或其他电活性化合物进行标记，将被测抗体或抗原的浓度信息加以化学放大，从而达到极高的灵敏度。

在电化学免疫传感器中抗体用于识别和特异性结合分析物到电极上，并用信号产生或放大分子标记结合的目标分析物。目前，电化学免疫传感器已广泛应用于医药与临床诊断和治疗[20-22]，例如诊断早期妊娠的 hCG 免疫传感器[23]、诊断原发性肝癌的甲胎蛋白（AFP 或 αFP）免疫传感器[24]、测定人血清蛋白（HSA）免疫传感器[25]、IgG 免疫传感器[26]、胰岛素免疫传感器等[27]。在电化学生物传感器中，由于各种原因，如增加电导率、电子转移、信号产生和放大，各种纳米材料被引入电极或标记分子中。因此，高效纳米材料的合成将提高电化学生物传感器的性能。总之，引入高效纳米材料或其复合偶联抗体的策略需要在未来的研究中进行评估。

1.1.2.4 组织电极与细胞器电极传感器

由于酶生物传感器的昂贵价格而且不稳定因此其应用受到限制，进而研发出了组织与细胞器电极传感器，开辟了生物传感器应用的新途径。直接采用动植物组织薄片作为敏感元件的电化学传感器称为组织电极传感器，其原理是利用动植物组织中的酶，优点是酶活性及其稳定性均比离析酶高，材料易于获取，制备简单，使用寿命长等。但在选择性、灵敏度、响应时间等方面还存在不足。动物组织电极主要有肾组织电极、肝组织电极、肠组织电极、肌肉组织电极、胸腺组织电极等。测定对象主要有谷氨酰胺、葡萄糖胺 6-磷酸盐、D-氨基酸、H_2O_2、地高辛、胰岛素、腺苷、AMP 等。植物组织电极敏感元件的选材范围很广，包括不同植物的根、茎、叶、花、果等。植物组织电极制备比动物组织电极更简单，成本更低并易于保存。

细胞器电极传感器是利用动植物细胞器作为敏感元件的传感器。细胞器是指存在于细胞内的被膜包围起来的微小"器官"，如线粒体、微粒体、溶酶体、过氧化氢体、叶绿体、氢化酶颗粒、磁粒体等，其原理是利用细胞器内所含的酶（多酶体系）。如利用人体内的红细胞过氧化氢酶的催化活性测定 H_2O_2 的传感器就是一种细胞器传感器。由动物癌细胞、酵母细胞或者细菌细胞固化制成的细胞传感器，可对抗癌药物和抗菌药物进行测定和筛选，并能监测各种活细胞新陈代谢的化学物质。

1.1.2.5 电化学 DNA 传感器

电化学 DNA 传感器是近几年迅速发展起来的一种全新思想的生物传感器[28]，其用途是检测基因及一些能与 DNA 发生特殊相互作用的物质。电化学 DNA 传感器是利用单链 DNA(ssDNA)或基因探针作为敏感元件固定在固体电极表面，加上识别杂交信息的电活性指示剂（杂交指示剂）共同构成的检测特定基因的装置，其工作原理是利用固定在电极表面的某一特定序列的 ssDNA 与溶液中的同源序列的特异识别作用（分子杂交）形成双链 DNA(dsDNA)（电极表面性质改变），同时借助一些能识别 ssDNA 和 dsDNA 的杂交指示剂的电流响应信号的改变来达到检测基因的目的。

有关 DNA 修饰电极的研究除对于基因检测有重要意义外[29,30]，还可将 DNA 修饰电极用于 DNA 与外源分子间的相互作用研究[31-33]，如抗癌药物筛选、抗癌药物作用机理研究。无疑，它将成为生物电化学的一个非常有生命力的前沿领域。然而，电化学 DNA 传感器还未能实现实际应用，主要是传感器的稳定性、重现性、灵敏度等都还有待于提高。

1.1.2.6 电化学非酶型纳米材料传感器

电化学酶传感器是以生物酶作为敏感元件的一类电化学传感器，其具有专一性强与检测限低等特点，然而生物酶的活性容易遭到破坏且种类有限致使应用受到制约[34]。随着纳米材料的快速发展及其展现出的独特物理与化学特性，使其在电池、电容、光电、机电、传感等领域得到了广泛应用[35-37]。纳米材料促进了电化学传感器研究取得了突破性的进展，其具有巨大的比表面积、良好的导电性能、高效的表面反应活性和较好的机械强度，并且由于其表面原子配位不相等使其拥有较多的表面活性位点、较强的催化能力和吸附能力[38-43]。此外，纳米材料的可塑性较强[44]，可以通过调控反应的温度、浓度、时间及配比来得到不同形貌、结合方式、电子状态等产物[45]。目前，常被用于电化学生物传感器的纳米材料主要包括纳米颗粒、纳米棒、纳米线、纳米管、纳米阵列、纳米笼等[46-49]。电化学非酶型纳米材料传感器常用于检测各种气体分子、金属离子、生物体内各种活性物质等。迄今为止，电化学非酶型纳米材料传感器在生物活性氧检测中的应用已被广泛研究与报道（见图 1.4）[50-52]，在此我们将主要对其相关研究进行概述。

非酶生物传感器的最新进展表明，非酶传感器面临的主要挑战是对特定分析物的选择性和准确测量分析物精确浓度的能力。尽管已经对能够检测生物活性物

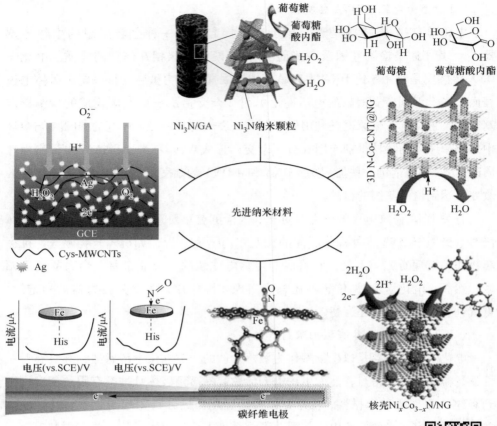

图 1.4　电化学非酶型纳米材料传感器

选择性电极纳米材料的开发进行了深入的研究，但仍有一些挑战需要解决：第一，缺乏临床实验室实际应用的纳米材料毒性研究；第二，需要了解分析物与宿主基质之间的相互作用，需要进行更基本的研究；第三，可重复性是另一个需要更多关注和研究的因素。由于纳米材料可以提供一个合适的生物相容环境，未来的研究可以指导设计体内生物传感器，以便在真实的生物系统中连续和长期监测目标分析物。此外，开发具有仿生酶活性的新型纳米材料将为检测生物分子开辟新的有效途径。因此，具有良好导电性和高表面积的功能化纳米材料将为构建适合实际应用的高灵敏度生物传感器提供良好的平台。

图 1.4 彩图

1.1.3　电化学生物传感器的表征方法

电化学生物传感器的表征方法是一种应用电化学原理来研究化学反应的方

法，其可以对电化学反应动力学、电化学过程及电化学界面性能等进行表征。为了研究电化学反应的特性以及电极材料的电化学性能，科学家们开发了多种表征方法。电化学生物传感器的性能表征方法采用传统的电化学测试方法，主要有以下几种。

(1) 循环伏安法（Cyclic Voltammetry，CV）是在工作电极上施加一个等腰三角形的脉冲电压，左半部分电位向阴极方向扫描，右半部分电位向阳极方向扫描，通过一次三角波扫描，完成一个循环[53]。循环伏安法可以用来研究电极反应的反应物浓度，电极表面吸附物的覆盖度，电极活性面积以及电极反应速率常数、交换电流密度和反应的传递系数等动力学参数。

(2) 计时电流（Current-Time，i-t）测试是对工作电极施加一个特定的电压，测量该电压下工作电极的电流密度随时间变化规律的电化学测试方法[54]。在电解水析氢反应中，只有电催化剂能够持续高效发生析氢反应，才具备实际的商业价值，否则尽管初始电催化性能极其优异，经过一段时间的测试衰减十分严重，也就没有商业应用价值。计时电位测试在电催化析氢领域评价电极材料的稳定性，设定一定的电压值，电极材料的电位在恒电压作用下，长时间进行计时电流测试，电流随时间的变化情况可反应电催化剂的稳定性。

(3) 电化学交流阻抗谱（Electrochemical Impedance Spectroscopy，EIS）[55,56]通常被用来研究电催化剂在 HER 过程的动力学信息。EIS 最常用的是阻抗复平面图和阻抗波特图。在 EIS 中，高-中频区表征是与化学反应的电荷传递阻抗有关，弧的直径越大表示电荷传递阻抗越大。此外，高频区的谱线与实轴的交点为系统的整体阻抗。一般中频区对应的是离子向电极扩散特征。低频区表示的是电解液中的离子向电极表面扩散时的难易程度。为了分析电化学反应过程中的机理，根据阻抗谱的响应建立电化学过程与等效电路之间的相关性。一般等效电路是由电阻、电容和电感等基本元件以串联或并联的方式组合，根据等效电路拟合出催化剂与溶液界面间的传输阻抗，可以评价催化剂的好坏。

(4) 线性扫描伏安法（Linear Sweep Voltammetry，LSV）是在工作电极与参比电极之间加上一个随时间进行线性变化的电极电势，同时记录通过工作电极与辅助电极之间的电流，从而获得电极电流与电极电位之间的伏安关系曲线，即线性扫描伏安图[57]。在电催化测试过程中，可以引入参比电极，电流不经过参比电极，无极化现象，电位稳定。LSV 方法应用于材料的电催化性能研究，是目前光电催化领域比较热门的测试手段之一。在电解水方面，通过线性扫描伏安曲线可以最直观地展示出电极的活性,在 LSV 曲线中,主要通过比较同一电流密度下的过电位大小来判断材料的电催化活性:过电位越小,催化活性越高。因此,在 LSV 测试过程中需要引入参比电极,构成三电极体系。在三电极测试体系中,从工作电极和参比电极得到电压信息,从工作电极和辅助电极得到电流信息,从而实现对工作电极电势和电流的准确测量。在三电极测试体系中,测试的电位相对于可逆氢电极的电位进行转换,其换算公式为:

$$ERHE = E_\circ + E + 0.059 \times pH \tag{1.1}$$

式中,$ERHE$ 为相对于可逆氢电极的电势;$E_\circ$ 为参比电极电势;E 为相对于参比电极电势。

在 LSV 测试中,可以得到析氢反应的动力学参数,包括 Tafel 斜率和交换电流密度。

电化学表征技术是一种通过测量反应中电流和电势随时间的变化来分析反应动力学和反应机理的技术。其在许多领域,如环境、材料科学等领域有广泛的应用前景。除了以上所述的应用方法外,还有数不胜数的技术和方法。未来,电化学表征技术将会在更多的领域中得到应用,并为各种问题提供解决方案。

1.1.4 电化学生物传感器的应用现状

生物传感器作为一种强大的创新分析工具,涉及生物传感元件和传感器,具有广泛的应用,如诊断、药物发现、生物医学、食品安全和加工、环境监测、安全和国防。生物技术、微电子和纳米技术领域的最新进展促进了生物传感器的发展,电化学生物传感器的发展前景十分广阔。随着技术的不断发展,它的小型化和高灵敏度将成为可能,同时,它的应用范围也将不断扩大。电化学生物传感器具有操作便捷、高灵敏度、高选择性等优点,在医学诊断方面,它可以帮助医生快速准确地诊断疾病,从而提高疾病的治愈率[58,59]。在环境监测方面,它可以实时监测水体中的有害物质,从而保护环境。在生物研究方面,它可以用于研究生物分子的相互作用,从而为药物研发提供基础数据[60]。研究将活性氧、氮物种和细胞色素 C 及其相关的酶或仿生酶(金属卟啉、纳米材料模拟酶)生物传感器应用于临床医学检验中,这些应用为临床诊断和病情的分析提供了重要参考依据。在生物医学、环境监测和食品安全监测等领域的应用越来越普遍。具体应用如下。

(1)生物医学:电化学生物传感器可以用于生物体液中蛋白检测(如肿瘤

标志物、特异性抗体)、小分子物质（如葡萄糖、抗生素、胆固醇、活性物质）、核酸（如病原微生物、突变基因）等多种生化指标的检测[61-66]。利用固定化葡萄糖氧化酶电极对氧或过氧化氢进行电化学测定的血糖仪为生物传感器的发现和发展奠定了基础。基于几何形状和相互作用力的分子识别在生物传感器的发展中起着重要作用。纳米技术的出现导致了高效和灵敏的生物传感器。它们还为各种生物受体提供了有效的固定基质。

（2）环境监测：传统的环境监测通常是采用离线分析方法，操作复杂，仪器昂贵，并且不适宜进行现场快速监测和连续在线分析。随着环境污染问题日益严重，电化学生物传感器在建立和发展连续、在线、快速的现场监测体系中发挥着重要作用。目前可以快速灵敏地监测水中重金属离子、三氯乙烯、硝酸盐、有机药物等污染物[67,68]。

（3）食品安全监测：随着社会经济的不断提高，目前人们对食品质量与安全性要求越来越重视，由于电化学生物传感器可以提供实时快速、简单便携和低成本的分析方法，电化学生物传感器能对食品中的农药残留（如对氧磷、百草枯等）[69,70]，以及食物中的成分进行快速分析，如蛋白质、糖类、有机酸、食品添加剂等[71,72]。

1.2 电化学生物传感器的发展趋势

随着生物技术和纳米技术的不断进步，开发用于监测生物分子（如蛋白质、药物、葡萄糖和代谢途径副产物）的敏感和选择性监测工具已成为新的趋势。目前，许多研究都集中在小型化和手持式传感设备的制造或开发，以期可用于生物分子特异性和敏感性检测的方法上。电化学技术有助于在所有分析技术中开发敏感、选择性、可重复和可再现的识别平台。除了提供所述的关键性能外，还可以使用电化学技术设计易于使用，价格合理，节省时间，环保和劳动友好的检测平台。在大多数应用中，研究人员已经充分利用电化学技术并将其应用于纳米技术领域，以获得小型化，更具选择性和灵敏度的工具作为检测平台。

癌症是导致肿瘤病人死亡的主要原因之一，通常使用抗肿瘤药物进行治疗，即使浓度很低，这些药物对人类和环境仍然具有较高的潜在毒性。因此，监测这些药物是至关重要的。在用于检测低浓度物质的技术中，电化学传感器和生物传感器以其实用性和低成本而闻名。电化学生物传感器在抗肿瘤药物定量分析中的发展和使用近年来越来越突出,并且使用电化学技术定量抗肿瘤药物可以带来巨大的经济和健康效益。随着电化学生物传感器在食品、医药、环境等领域应用范围的不断扩大,对传感器的性能提出了更高的要求,虽然从理论上研究电化学生物传感器具有一般的电化学传感器所难以比拟的优越性,特别是在选择性和灵敏度方面,但由于生物分子的引入,生物分子结构本身固有的不稳定性与易变性,使得生物传感器在实际应用中还存在较多的问题。因此,为了研制出高灵敏度、高稳定性、低成本的电化学生物传感器,必须克服生物单元结构的易变性,使其稳定性增加,最常用的手段是使用具有生物单元稳定作用的介质。其次,对一些特定的分析对象提高电信号,降低检测限。目前,电化学生物传感器在癌症药物的研制方面正产生着不小的作用。有研究者实现了对癌细胞的体外培养,并通过电化学生物传感器准确地测试癌细胞对各种治癌药物的反应,经过一系列研究性试验可快速筛选出拥有最佳反应药效的治癌药物[60,73]。

在当今科技高速发展的时代,纳米材料科学的迅速发展使得不同尺寸、材质和大小的纳米材料已被普遍应用在生物传感器中,与传统的材料相比,纳米材料具有无可比拟的优越性:(1)纳米级尺寸使其具有大比表面积;(2)良好的生物相容性;(3)结构性能稳定,室温下长时间保存不影响材料的性能;(4)化学与物理性质易于控制。纳米材料具备的这些特点,使其被广泛用于构建高特异性和高灵敏度的生物小分子传感器[74]。Guo 等人[75]报道了石墨烯/人工过氧化氢酶/蛋白质构建的层状多功能生物界面用于细胞释放的 H_2O_2 的原位定量检测。此传感器具有 4.5 $\mu A/(\mu mol \cdot L^{-1} \cdot cm^2)$ 的灵敏度,检测限可以达到 0.1×10^{-6} mol/L (S/N=3),响应时间小于 5 s。纳米多功能电极提供了有益于细胞的三维空间生长环境,这项工作将有可能促使石墨烯及其层状结构具有较高的功能可控性和可调性,可用于原位细胞的测定和下一代智能设备的探索,例如,开发药物释放可控系统,分子电子系统以及高能量存储的便携式设备。Yu 等人[76]报道了聚电解质金纳米颗粒复合膜修饰导电玻璃(ITO)电极检测 NO。该电极在 0.82 V 附近出现一个较明显的氧化峰,由于金纳米粒子催化 NO 的氧化反应,随着 $NaNO_2$ 浓度的增加,NO 的氧化峰电流也在不断增加。因此,金纳米颗粒修饰的电极可以通过电极表面的改性、自组装等用于 NO 的检测。由于一氧化氮与超氧阴离子在医学与生物学方面具有重要的作用,研究二者的定量检测方法对于理解活性氧自由基相关疾病的病理学与生物学起着至关重要的作用。因此,课题组应用纳米材料构建了活性氧纳米复合材料电化学传感器。

1.3 电化学活性氧和活性氮传感器概述

细胞活性氧和活性氮（Reactive Oxygen and Nitrogen Species, RONS）是生物系统中作为代谢副产物生成的高活性分子，也是细胞信号分子，对机体的健康具有非常重要的作用。为了能够更深入了解活性氧在病理学和生理学中的重要作用，研究出一种高效、准确、专一性强的活性氧的检测方法显得十分必要。传统的检测活性氧的方法主要有电子自旋共振法、分光光度法、高效液相色谱法、化学发光法、荧光法等。但由于细胞活性氧和活性氮在生物体内存在浓度低、反应活性高、寿命短等的限制，活性氧在生理环境下的检测仍然是一个巨大的挑战。然而应用纳米复合材料制备的活性氧电化学传感器，利用电化学方法实现对活性氧进行快速分析，其所需仪器简单、选择性好、分析速度快、检测成本低、易于实现微型化等优点，在生物医学、环境监测、食品和医药[77,78]等领域具有广泛的应用前景，这种分析技术终将替代传统的活性氧检测方法。

生物细胞活性氧和活性氮传感器是一类分析器件，它将生物活体细胞释放的细胞活性氧和活性氮小分子作为研究对象，采用物理化学传感器或传感微系统如光学或电化学信号元件，实时、定性并定量地检测目标分析物。活性氧小分子传感器主要包括信号识别产生系统、信号转换的换能系统和信号输出系统（见图1.1）。当活细胞受到药物刺激释放出小分子后，该小分子与传感界面产生特异性反应，发出的信息变化通过换能器转换为可处理的间断或连续的数字信号，从而实现定性、定量地检测活细胞释放细胞活性氧和活性氮小分子的性质或含量。

1.4 小　结

近年来，随着科学技术的飞速发展癌症的防治与治疗也在一定程度上得到了显著的提升。通常情况下，如果能及时采取正确的预防措施，目前1/3的癌症是可以预防的，因此，肿瘤早期的诊断与预防是降低肿瘤对人类危害的重要措施之一。目前已有报道活性氧的过度释放能够导致细胞氧化损伤，进而引起蛋白质、糖类和DNA的结构和生理功能发生变化，最终导致细胞凋亡[79,80]。多项研究已表明氧化损伤是引起许多组织细胞凋亡的一个重要机制[81]。相关研究药物和环境的外界刺激对细胞的氧化损伤已经证明了活性氧分子与细胞氧化损伤有密切的关系，但是目前的研究主要集中在基因的变化、细胞蛋白质和酶的破坏及线粒体的损伤，对应用电化学传感器分析细胞氧化损伤与活性氧分子浓度变化关系的研究较少。由于活性氧分子的半衰期短，细胞释放的活性氧浓度低和活性高等特点，导致活性氧的检测相当困难。传统的生物学方法可以间接评估细胞释放的活性氧，这种检测方法具有耗时、样品需要量大、需要一定的标记物进行标记、费用昂贵等不足。因此，研制出一种快速、灵敏、便捷的方法应用于实时、原位检测生物活性氧对研究人类的生理功能和疾病早期诊断具有重要意义。

众所周知，RONS 是与不同病理相关的氧化应激标志物，包括神经退行性疾病、心血管疾病以及癌症。因此，生物系统中 RONS 水平的检测是生物医学和分析研究的一个重要问题。在此，主要研究通过对非酶标记的电化学生物传感器监测肿瘤细胞中 RONS 的研究进行综述与概括，希望为抗肿瘤药物的开发、疾病早期诊断与治疗提供理论参考依据。

参 考 文 献

[1] THEVENOT D R, TOTH K, DURST R A, et al. Electrochemical biosensors: Recommended definitions and classification [J]. Biosensors & Bioelectronics, 2001, 16 (1/2): 121-131.

[2] CESEWSKI E, JOHNSON B N. Electrochemical biosensors for pathogen detection [J]. Biosens Bioelectron, 2020, 159: 112214.

[3] KAYA H O, CETIN A E, AZIMZADEH M, et al. Pathogen detection with electrochemical biosensors: Advantages, challenges and future perspectives [J]. J Electroanal Chem (Lausanne), 2021, 882: 114989.

[4] HASSAN R Y A. Advances in electrochemical nano-biosensors for biomedical and environmental applications: From current work to future perspectives [J]. Sensors (Basel), 2022, 22 (19): 7539.

[5] ZOLTI O, SUGANTHAN B, RAMASAMY R P. Lab-on-a-chip electrochemical biosensors for foodborne pathogen detection: A review of common standards and recent progress [J]. Biosensors (Basel), 2023, 13 (2): 215.

[6] ZHU Z, SONG H, WANG Y, et al. Protein engineering for electrochemical biosensors [J]. Curr Opin Biotechnol, 2022, 76: 102751.

[7] ZHAI Q, CHENG W. Soft and stretchable electrochemical biosensors [J]. Mater Today Nano, 2019, 7: 100041.

[8] YOON J, CHO H Y, SHIN M, et al. Flexible electrochemical biosensors for healthcare

monitoring [J]. J Mater Chem B, 2020, 8 (33): 7303-7318.

[9] SHAN C, YANG H, SONG J, et al. Direct electrochemistry of glucose oxidase and biosensing for glucose based on graphene [J]. Analytical Chemistry, 2009, 81 (6): 2378-2382.

[10] DU J, YU X, DI J. Comparison of the direct electrochemistry of glucose oxidase immobilized on the surface of Au, CdS and ZnS nanostructures [J]. Biosensors and Bioelectronics, 2012, 37 (1): 88-93.

[11] LEE H, HONG Y J, BAIK S, et al. Enzyme-based glucose sensor: From invasive to wearable device [J]. Advanced Healthcare Materials, 2018, 7 (8): 1701150.

[12] ZHAO L, WEN Z, JIANG F, et al. Silk/polyols/GOD microneedle based electrochemical biosensor for continuous glucose monitoring [J]. Rsc Adv, 2020, 10 (11): 6163-6171.

[13] TIMUR S, ANIK U, ODACI D, et al. Development of a microbial biosensor based on carbon nanotube (CNT) modified electrodes [J]. Electrochem Commun, 2007, 9 (7): 1810-1815.

[14] HAN X J, LI C, YONG D M. Microbial electrode sensor for heavy-metal ions [J]. Sensor Mater, 2019, 31 (12): 4103-4111.

[15] GRUNDIG B, KRABISCH C. Electron mediator-modified electrode for the determination of glucose in fermentation media [J]. Anal Chim Acta, 1989, 222 (1): 75-81.

[16] HUANG Y L, FOELLMER T J, ANG K C, et al. Characterization and application of an on-line flow injection analysis wall-jet electrode system for glucose monitoring during fermentation [J]. Anal Chim Acta 1995, 317 (1/2/3): 223-232.

[17] APPELQVIST R, HANSEN E H. Determination of glucose in fermentation processes by means of an on-line coupled flow-injection system using enzyme sensors based on chemically modified electrodes [J]. Anal Chim Acta, 1990, 235: 265-271.

[18] KARUBE I, WANG Y X, TAMIYA E, et al. Microbial electrode sensor for vitamin-B12 [J]. Anal Chim Acta, 1987, 199: 93-97.

[19] PIRO B, REISBERG S. Recent advances in electrochemical immunosensors [J]. Sensors, 2017, 17 (4): 794.

[20] WEN W, YAN X, ZHU C Z, et al. Recent advances in electrochemical immunosensors [J]. Analytical Chemistry, 2017, 89 (1): 138-156.

[21] MOLLARASOULI F, KURBANOGLU S, OZKAN S A. The role of electrochemical immunosensors in clinical analysis [J]. Biosensors-Basel, 2019, 9 (3): 86.

[22] POLLAP A, KOCHANA J. Electrochemical immunosensors for antibiotic detection [J]. Biosensors-Basel, 2019, 9 (2): 61.

[23] LU W, CHEN Z A, WEI M, et al. A three-dimensional CoNi-MOF nanosheet array-based immunosensor for sensitive monitoring of human chorionic gonadotropin with core-shell ZnNi-

MOF@ Nile Blue nanotags [J]. Analyst, 2020, 145 (24): 8097-8103.

[24] IDRIS A O, MABUBA N, AROTIBA O A. An alpha-fetoprotein electrochemical immunosensor based on a carbon/gold bi-nanoparticle platform [J]. Anal Methods-Uk, 2018, 10 (47): 5649-5658.

[25] STANKOVIC V, DURDIC S, OGNJANOVIC M, et al. Anti-human albumin monoclonal antibody immobilized on EDC-NHS functionalized carboxylic graphene/AuNPs composite as promising electrochemical HSA immunosensor [J]. Journal of Electroanalytical Chemistry, 2020, 860: 113928.

[26] GOODE J, DILLON G, MILLNER P A. The development and optimisation of nanobody based electrochemical immunosensors for IgG [J]. Sensors and Actuators B-Chemical, 2016, 234: 478-484.

[27] ZARE Y, SOLEYMANI J, RAHIMI M, et al. Trends in advanced materials for the fabrication of insulin electrochemical immunosensors [J]. Chem Pap, 2022, 76 (12): 7263-7274.

[28] HAI X, LI Y F, ZHU C Z, et al. DNA-based label-free electrochemical biosensors: From principles to applications [J]. TrAC-Trends in Analytical Chemistry, 2020, 133: 116098.

[29] NAGRAIK R, SETHI S, SHARMA A, et al. Ultrasensitive nanohybrid electrochemical sensor to detect gene of [J]. Chem Pap, 2021, 75 (10): 5453-5462.

[30] EVTUGYN G A, PORFIREVA A V, BELYAKOVA S V. Electrochemical DNA sensors for drug determination [J]. J Pharmaceut Biomed, 2022, 221: 115058.

[31] KULIKOVA T, PORFIREVA A, EVTUGYN G, et al. Electrochemical DNA sensors with layered polyaniline-DNA coating for detection of specific DNA interactions [J]. Sensors, 2019, 19 (3): 469.

[32] DINÇER E, KÜÇÜKOGLU N, KIVANÇ M, et al. Electrochemical DNA sensor designed using the pencil graphite electrode to detect [J]. Appl Biochem Biotech, 2024, 196 (7): 4679-4698.

[33] WETTASINGHE A P, SINGH N, STARCHER C L, et al. Detecting attomolar DNA-damaging anticancer drug activity in cell lysates with electrochemical DNA devices [J]. ACS Sensors, 2021, 6 (7): 2622-2629.

[34] LIU J Z, LI M Z, LIU W D, et al. Advances in non-enzymatic electrochemical materials for H_2O_2 sensing [J]. Journal of Electroanalytical Chemistry, 2024, 954: 118060.

[35] SEIFI H, GHOLAMI T, SEIFI S, et al. A review on current trends in thermal analysis and hyphenated techniques in the investigation of physical, mechanical and chemical properties of nanomaterials [J]. Journal of Analytical and Applied Pyrolysis, 2020, 149: 104840.

[36] LI Z, WANG L, LI Y, et al. Carbon-based functional nanomaterials: Preparation, properties

and applications [J]. Composites Science and Technology, 2019, 179: 10-40.

[37] CHEN Y, FAN Z, ZHANG Z, et al. Two-dimensional metal nanomaterials: Synthesis, properties, and applications [J]. Chemical Reviews, 2018, 118 (13): 6409-6455.

[38] ZHAO W, CHEN I W, HUANG F. Toward large-scale water treatment using nanomaterials [J]. Nano Today, 2019, 27: 11-27.

[39] SALARIZADEH P, ASKARI M B, MOHAMMADI M, et al. Electrocatalytic performance of CeO_2-decorated rGO as an anode electrocatalyst for the methanol oxidation reaction [J]. Journal of Physics and Chemistry of Solids, 2020, 142: 109442.

[40] XIANG G, WANG Y G. Exploring electronic-level principles how size reduction enhances nanomaterial surface reactivity through experimental probing and mathematical modeling [J]. Nano Research, 2022, 15 (4): 3812-3817.

[41] JEON S K, JANG H S, KWON O H, et al. Mechanical test method and properties of a carbon nanomaterial with a high aspect ratio [J]. Nano Convergence, 2016, 3 (1): 29.

[42] NI B, WANG X. Face the edges: Catalytic active sites of nanomaterials [J]. Advanced Science, 2015, 2 (7): 1500085.

[43] WEI Y, LIU X, WANG Z, et al. Adsorption and catalytic degradation of preservative parabens by graphene-family nanomaterials [J]. Science of The Total Environment, 2022, 806: 150520.

[44] GUO W L, GUO Y F, ZHANG Z H, et al. Strength, plasticity, interlayer interactions and phase transition of low-dimensional nanomaterials under multiple fields [J]. Acta Mech Solida Sin, 2012, 25 (3): 221-243.

[45] SALEH T A. Nanomaterials: Classification, properties, and environmental toxicities [J]. Environmental Technology & Innovation, 2020, 20: 101067.

[46] LI Z, LIU C, SARPONG V, et al. Multisegment nanowire/nanoparticle hybrid arrays as electrochemical biosensors for simultaneous detection of antibiotics [J]. Biosensors and Bioelectronics, 2019, 126: 632-639.

[47] AMEEN S, AKHTAR M S, SHIN H S. Nanocages-augmented aligned polyaniline nanowires as unique platform for electrochemical non-enzymatic glucose biosensor [J]. Applied Catalysis A: General, 2016, 517: 21-29.

[48] KUMAR B, SINHA S K. Nanostructured Cu_2O deposited on TiO_2 nanotube arrays for ultra-sensitive non-enzymatic glucose electrochemical biosensor [J]. Ionics, 2023, 29 (2): 793-805.

[49] MAZAHERI M, AASHURI H, SIMCHI A. Three-dimensional hybrid graphene/nickel electrodes on zinc oxide nanorod arrays as non-enzymatic glucose biosensors [J]. Sensors and

Actuators B: Chemical, 2017, 251: 462-471.

[50] THATIKAYALA D, PONNAMMA D, SADASIVUNI K K, et al. Progress of advanced nanomaterials in the non-enzymatic electrochemical sensing of glucose and H_2O_2 [J]. Biosensors (Basel), 2020, 10 (11): 151.

[51] GUO J, LI M, LONG S, et al. Bio-inspired electrochemical detection of nitric oxide promoted by coordinating the histamine-iron phthalocyanine catalytic center on microelectrode [J]. Analytical Chemistry, 2023, 95 (23): 8842-8849.

[52] LIU Y, LIU X, LIU Y, et al. Construction of a highly sensitive non-enzymatic sensor for superoxide anion radical detection from living cells [J]. Biosensors and Bioelectronics, 2017, 90: 39-45.

[53] ELGRISHI N, ROUNTREE K J, MCCARTHY B D, et al. A practical beginner's guide to cyclic voltammetry [J]. Journal of Chemical Education, 2018, 95 (2): 197-206.

[54] THORGAARD S N, JENKINS S, TARACH A R. Influence of electroosmotic flow on stochastic collisions at ultramicroelectrodes [J]. Analytical Chemistry, 2020, 92 (18): 12663-12669.

[55] WANG S, ZHANG J, GHARBI O, et al. Electrochemical impedance spectroscopy [J]. Nature Reviews Methods Primers, 2021, 1 (1): 41.

[56] PAJKOSSY T, JURCZAKOWSKI R. Electrochemical impedance spectroscopy in interfacial studies [J]. Curr Opin Electroche, 2017, 1 (1): 53-58.

[57] ATOURKI L, IHALANE E H, KIROU H, et al. Characterization of nanostructured ZnO grown by linear sweep voltammetry [J]. Solar Energy Materials and Solar Cells, 2016, 148: 20-24.

[58] HATAMIE A, HE X, ZHANG X W, et al. Advances in nano/microscale electrochemical sensors and biosensors for analysis of single vesicles, a key nanoscale organelle in cellular communication [J]. Biosens Bioelectron, 2023, 220: 114899.

[59] SADEGHI M, SADEGHI S, NAGHIB S M, et al. A comprehensive review on electrochemical nano biosensors for precise detection of blood-based oncomarkers in breast cancer [J]. Biosensors (Basel), 2023, 13 (4): 481.

[60] LIMA H R S, DA SILVA J S, DE OLIVEIRA FARIAS E A, et al. Electrochemical sensors and biosensors for the analysis of antineoplastic drugs [J]. Biosens Bioelectron, 2018, 108: 27-37.

[61] ANDREESCU S, VASILESCU A. Advances in electrochemical detection for probing protein aggregation [J]. Curr Opin Electroche, 2021, 30: 100820.

[62] LI D J, WEI H Y, HONG R, et al. WS_2 nanosheets-based electrochemical biosensor for highly sensitive detection of tumor marker miRNA-4484 [J]. Talanta, 2024, 274: 125965.

[63] AHMAD M, NISAR A, SUN H Y. Emerging trends in non-enzymatic cholesterol biosensors:

Challenges and advancements [J]. Biosensors-Basel, 2022, 12 (11): 955.

[64] SHOJA Y, KERMANPUR A, KARIMZADEH F, et al. Electrochemical molecularly bioimprinted siloxane biosensor on the basis of core/shell silver nanoparticles/EGFR exon 21 L858R point mutant gene/siloxane film for ultra-sensing of gemcitabine as a lung cancer chemotherapy medication [J]. Biosensors & Bioelectronics, 2019, 145: 111611.

[65] GONG C C, SHEN Y, SONG Y H, et al. On-off ratiometric electrochemical biosensor for accurate detection of glucose [J]. Electrochim Acta, 2017, 235: 488-494.

[66] BRACAGLIA S, RANALLO S, RICCI F. Electrochemical cell-free biosensors for antibody detection [J]. Angew Chem Int Edit, 2023, 62 (8): e20226512.

[67] KUKLA A L, KANJUK N I, STARODUB N F, et al. Multienzyme electrochemical sensor array for determination of heavy metal ions [J]. Sensors and Actuators B-Chemical, 1999, 57 (1/2/3): 213-218.

[68] MOSCHOU E A, LASARTE U A, FOUSKAKI M, et al. Direct electrochemical flow analysis system for simultaneous monitoring of total ammonia and nitrite in seawater [J]. Aquacult Eng, 2000, 22 (4): 255-268.

[69] ALVAREZ M, CALLE A, TAMAYO J, et al. Development of nanomechanical biosensors for detection of the pesticide DDT [J]. Biosensors & Bioelectronics, 2003, 18 (5/6): 649-653.

[70] JIANG X S, LI D Y, XU X, et al. Immunosensors for detection of pesticide residues [J]. Biosensors & Bioelectronics, 2008, 23 (11): 1577-1587.

[71] MORALES M D, GONZALEZ M C, REVIEJO A J, et al. A composite amperometric tyrosinase biosensor for the determination of the additive propyl gallate in foodstuffs [J]. Microchem J, 2005, 80 (1): 71-78.

[72] DAS J, JO K, LEE J W, et al. Electrochemical immunosensor using p-aminophenol redox cycling by hydrazine combined with a low background current [J]. Analytical Chemistry, 2007, 79 (7): 2790-2796.

[73] HASSANI MOGHADAM F, TAHER M A, KARIMI-MALEH H. Doxorubicin anticancer drug monitoring by ds-DNA-based electrochemical biosensor in clinical samples [J]. Micromachines (Basel), 2021, 12 (7): 808.

[74] HUANG X P, ZHU Y F, KIANFAR E. Nano biosensors: Properties, applications and electrochemical techniques [J]. J Mater Res Technol, 2021, 12: 1649-1672.

[75] GUO C X, ZHENG X T, LU Z S, et al. Biointerface by cell growth on layered graphene-artificial peroxidase-protein nanostructure for in situ quantitative molecular detection [J]. Adv Mater, 2010, 22 (45): 5164.

[76] YU A M, LIANG Z J, CHO J, et al. Nanostructured electrochemical sensor based on dense

gold nanoparticle films [J]. Nano Lett, 2003, 3 (9): 1203-1207.
[77] YU D H, BLANKERT B, VIRE J C, et al. Biosensors in drug discovery and drug analysis [J]. Anal Lett, 2005, 38 (11): 1687-1701.
[78] BERTUCCI C, CIMITAN S. Rapid screening of small ligand affinity to human serum albumin by an optical biosensor [J]. J Pharmaceut Biomed, 2003, 32 (4/5): 707-714.
[79] NIEMIETZ I, BROWN K L. Hyaluronan promotes intracellular ROS production and apoptosis in TNFα-stimulated neutrophils [J]. Frontiers in immunology, 2023, 14: 1032469.
[80] JIANG X, LI G, ZHU B, et al. P20BAP31 induces cell apoptosis via both AIF caspase-independent and the ROS/JNK mitochondrial pathway in colorectal cancer [J]. Cellular & molecular biology letters, 2023, 28 (1): 25.
[81] CRUZ-GREGORIO A, ARANDA-RIVERA A K, APARICIO-TREJO O E, et al. Alpha-Mangostin induces oxidative damage, mitochondrial dysfunction, and apoptosis in a triple-negative breast cancer model [J]. Phytother Res, 2023, 37 (8): 3394-3407.

2 纳米酶电化学生物传感器在活性氧和活性氮检测中的应用

2.1 引　言

　　细胞活性氧和活性氮是指在生物体内与氧代谢有关的含氧或含氮自由基和易形成自由基的过氧化物的总称[1]。机体内氧化代谢可不断形成活性氧，在一定的空间、时间和一定的限度内活性氧有积极的生理作用。它们是活性物质家族的一部分，包括活性氮、硫、碳、硒、亲电试剂和卤素（RHS）物质，它们可以进行氧化还原（还原-氧化）反应并在生物大分子上形成氧化修饰，从而有助于氧化还原信号传导和生物功能。然而，目前研究已经明确，高于生理浓度的 RONS 与蛋白质、脂质、核酸和碳水化合物发生非特异性反应，并产生其他具有潜在毒性后果的反应物质[2]。细胞应激反应系统通过感知不同细胞中氧化剂水平偏离稳态设定值的情况，进而启动适当的对抗措施，维持体内平衡并防止这种损害[3,4]。目前对于生理（有益）氧化应激（Eustress）和超生理（有害）氧化应激（Distress）的调控机制一直在不断地探索，同时对于氧化应激反应系统如何控制细胞氧化还原能力也在继续研究。在生理氧化应激中，氧化剂以低水平存在，并与生理氧化还原信号的特定靶标发生反应，而在氧化应激中，当氧化剂与非特异性靶标发生反应时，氧化剂浓度的增加会导致氧化还原信号异常或中断[5]。由于现有大量 RONS 相关文献，我们主要关注超氧阴离子自由基（$O_2^{·-}$）、一氧化氮（NO）和过氧化氢（H_2O_2）论述对象讨论其在肿瘤研究中的相关应用。

2.2 细胞活性氧和活性氮的生物学功能

　　活性氧（Reactive Oxygen Species，ROS）于 1954 年在生物环境中被发现[6]。从那时起，活性氧一直是研究的重点，研究它们的化学和生理活性，以及它们在生物体中的病理作用[7]。ROS是自由基的总称，自由基是指含有一个或多个氧的不成对电子的物质，如羟基（·OH）和超氧化物（O_2^-），以及容易转化为自由基的非自由基氧化剂，如过氧化氢（H_2O_2）和次氯酸（HOCl）[8]。活性氧和氮种（RONS）一词用于进一步涵盖这类含有氮或活性氮种（RNS）的化学反应分子，如一氧化氮（NO·）和过氧亚硝酸盐（$ONOO^-$）。一般来说，活性氮（Reactive Nitrogen Species, RON）寿命很短，在微秒或纳秒量级，并且很容易与包括脂质、核酸和蛋白质在内的许多细胞成分发生反应。这一过程通过自由基链反应发生，造成损伤并形成有害的二次产物，如脂质过氧化物和其他脂质加合物。RON信号的紊乱会导致它们的过量产生或解毒水平降低，从而导致氧化还原状态的改变，进而导致细胞成分的氧化损伤。RONS的存在与多种疾病有关，如中风[9]、败血症[10]、糖尿病[11]、高血压[12]、神经退行性疾病[13]、炎症[14]和癌症[15]。RONS在生物系统生理和病理过程中具有的双重作用已被广泛报道。最近的研究表明，在生理和病理条件下调节RONS水平分别是促进健康和治疗疾病的潜在疗法。因此，选择性调控它们在细胞内的产生或靶向它们的失活，可能为治疗ROND相关疾病提供了一种新的治疗策略。

2.2.1　细胞中活性氧和活性氮（RONS）概述

活性物种或自由基，包括活性氧和活性氮是生物系统中作为代谢副产物生成的高活性分子，也是细胞信号分子。活性氧（ROS）是指各种化学衍生的氧分子，包括自由基，如超氧化物（O_2^-）、羟基（·OH）、过氧基（RO_2）和烷氧基（RO），以及非自由基，如次氯酸（HOCl）、臭氧（O_3）、单线态氧（$1O_2$）和过氧化氢（H_2O_2）。在这些ROS中，H_2O_2是一种稳定的非自由基氧化物，作为细胞内信号分子能在生理含量水平较低时调节激酶驱动的途径。ROS的胞内来源包括线粒体、内质网、细胞核、质膜、过氧化物酶体甚至细胞外间隙。许多细胞酶能催化释放ROS，如NO合酶、过氧化物酶、NADPH氧化酶、NADPH氧化酶异构体和葡萄糖氧化酶。

与之相反，一氧化氮（NO）是由一氧化氮合成酶（NOS）催化L-精氨酸所生成的主要活性氮（RNS）分子。RNS的另一种活性形式是过氧亚硝酸根（$ONOO^-$），它由·NO和O_2^-在低氧张力条件下经相互作用形成。这些RONS在健康和疾病中都发挥着重要的生理和病理作用。在过去，高RONS水平长期被看作氧化应激，并与许多疾病的发展密切相关，如帕金森病和阿尔茨海默病、心血管疾病、癌症、糖尿病和类风湿性关节炎。然而，在生理水平上，RONS作为细胞内信号级联的重要氧化还原信使来发挥作用。例如，在伤口愈合过程中，它们能调节胶原形成，细胞增殖、分化和迁移。RONS 还通过调节血压，抑制血小板黏附和激活、防止平滑肌细胞增殖来保护心脏。在先天性免疫系统中，RONS 能对抗病原体保护宿主。因此，了解 RONS 的作用机制，并精确控制它们在生物环境中的含量，将为治疗各种与 RONS 相关的疾病提供新的和有潜力的疗法。自由基是含有一个或多个未配对电子的分子/分子片段，这些电子的存在通常使它们具有高活性[16]。近年来，活性氧的检测在医学、生物学和化学等领域受到了高度的重视。活性氧是一切需氧生物新陈代谢过程中的必然产物。由于活性氧自由基在机体内信号调控的研究越来越深入，其在生物学中的作用受到了人们极大的关注。为此，建立体外活性氧的实验检测平台，通过研究活性氧诱导的氧化损伤来进行药物的筛选就显得尤为重要。活性氧物质的形成主要是通过电子转移反应、能量转移反应和电离反应等途径来实现的。

在人体内，最常见的活性氧氮物质是超氧自由基、单线态氧、高氯酸根离子等，它们的生成多与细胞呼吸代谢及外界刺激等因素有关。RONS 通常包括超氧阴离子（$O_2^{·-}$ 和 $HO^{·-}$）、氢过氧自由基（$HO_2^{·-}$）、羟基自由基（$HO·$）、过氧化自由基（$RO_2·$）、烷氧基（$RO·$）等自由基以及某些易于转化为自由基的非自由基物质，如单线态氧（$1O_2$）、过氧化氢（H_2O_2）、一氧化氮（NO）及其衍生物等，其分子排布方式如图 2.1 所示[17]。

图 2.1 彩图

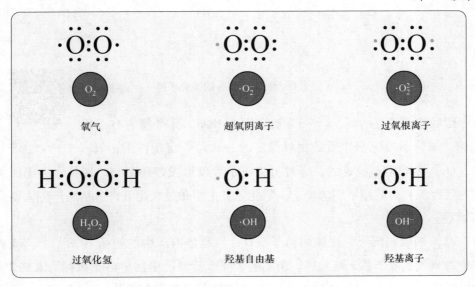

图 2.1　活性氧自由基的分子结构式

机体内氧化代谢可不断形成活性氧，在一定的空间、时间和一定的限度内活性氧有积极的生理作用，但是一旦过量产生，就会对身体健康产生严重的负面影响。需氧型生物进行新陈代谢都离不开 ROS。线粒体是产生活性氧的最重要场所[18]，产生活性氧其他场所如过氧化物酶体、细胞膜、内质网、核膜和核骨架，主要由线粒体内的电子传递链以及质膜上的 NADPH 氧化酶生成（见图 2.2）。

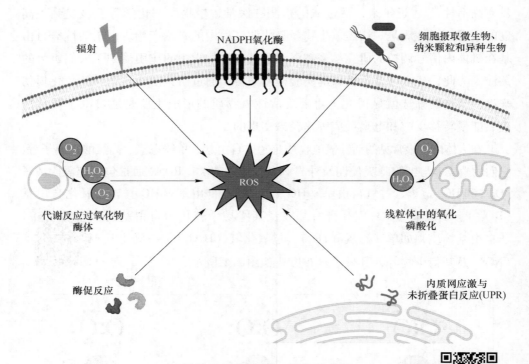

图 2.2　细胞中 ROS 的主要来源

线粒体中消耗的氧 1%~5% 被还原为 ROS，其产物为 $O_2^{·-}$ 和 H_2O_2。氧分子的化学性质是相对稳定的，然而一定条件下，氧分子（O_2）是非常不稳定的，通过得失电子可以形成各种 ROS。现将本书中所关注的活性氧自由基超氧阴离子、一氧化氮自由基和过氧化氢自由基的产生简要介绍如下。

（1）超氧阴离子。超氧阴离子（$O_2^{·-}$）是体内产生的初始 ROS[19,20]。线粒体中的氧气主要来源于线粒体，由于电子呼吸传递链中的单电子泄漏，大约 2% 的线粒体氧气被还原为 $O_2^{·-}$[21]。$O_2^{·-}$ 经过酶促或非酶促催化，形成各种对生理功能至关重要的活性氧[22,23]。如图 2.3 所示，在超氧化物歧化酶（SOD）的快速自猝灭作用下，$O_2^{\cdot-}$ 转化为相对稳定的 H_2O_2。这种转化导致两种结果：通过碳酸酐酶（CAT）或谷胱甘肽过氧化物酶（GPX）产生水分子，以及在 Fe^{2+}/Cu^+ 离子存在下启动芬顿反应，产生高活性的 $\cdot OH$。此外，H_2O_2 在髓过氧化物酶（MPO）的催化下生成 HOCl。$O_2^{\cdot-}$ 也与一氧化氮自由基（NO·）发生非酶性反应，生成 $ONOO^-$，$ONOO^-$ 可进一步分解为 NO 和 $\cdot OH$[24]。因此，监测 $O_2^{\cdot-}$ 水平可以在一定程度上作为评估其他 ROS 水平变化的指标。

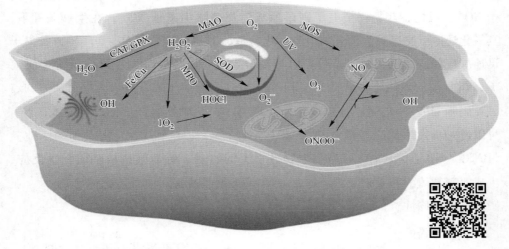

图 2.3　活细胞中各种活性氧的产生和转化示意图[25]　　图 2.3 彩图

$O_2^{\cdot-}$ 是生物系统中一种重要的 ROS，在细胞功能调节中起着广泛的作用，与许多疾病的发生和发展密切相关。超氧阴离子（$O_2^{\cdot-}$）是一种含自由基阴离子的活性氧，自由基态氧分子获得一个电子所得的活性中间体，在无催化剂或生物酶时，其自身可以发生歧化反应。它是生物免疫系统的重要成分，在细胞中具有重要的生物功能（见图 2.3）。超氧阴离子在水溶液中的存活时间是 1 s，在脂溶液中的存活时间是 1 h，与其他的 ROS 自由基相比半衰期较长，$O_2^{\cdot-}$ 虽然不是很活泼，但由于它的寿命相对比较长，可以扩散到较远距离进行反应，从这方面意义上说，$O_2^{\cdot-}$ 具有较大的危险性，如图 2.3 所示[26]。

在机体的新陈代谢过程中，一些酶可以催化还原反应的发生，使 O_2 还原为 $O_2^{\cdot-}$。例如，DADPH-细胞色素 CP450 还原酶就可以催化 DADPH 拿出一个 e，使细胞色素 CP450 中含有的血红素-Fe^{3+} 得到一个 e 生成血红素-Fe^{2+}，有氧气时后者又生成血红素-Fe^{2+}-O_2 中间体,最终生成血红素-Fe^{3+}-$O_2^{·-}$ [27]。此外,在电化学测试过程中通常采用超氧化钾(KO_2)的分解产生 $O_2^{·-}$,其反应如式(2.1)所示。

$$KO_2 \longrightarrow K^+ + O_2^{·-} \quad (2.1)$$

(2)一氧化氮。一氧化氮的 N 原子含有一个孤立电子,是典型的自由基,因而具有顺磁性。一氧化氮相对分子质量为 30,是非常活泼的分子,它的半衰期短仅有 5~10 s,具有较高的脂溶性,因此它可以穿过细胞膜进入细胞,并扩散到周围的组织细胞发出交流信号。由于一氧化氮化学性质活泼,主要发生如下四种反应(见图 2.4)。1)通过 NO 的孤立电子给出电子,氧化成为亚硝酰阳离子 NO^+;反之,获得一个电子则还原成为亚硝酰阴离子 NO^-,在生理条件下,其氧化还原产物已发生消化胁迫与氧化胁迫,对机体造成损伤。2)与金属离子迅速反应如铁、铜、镁等,还可以与高价含氧金属离子或发生反应低价金属离子,如 Fe^{4+} 的氧化还原反应[28]。3)NO 与 $O_2^{·-}$ 反应生成具有较大氧化能力的过氧亚硝基阴离子 $ONOO^-$。4)与 O_2 生成 NO_2,它对细胞产生的毒性更大。NO_2 可以进一步发生反应生成 N_2O_3、NO、NO_2^-[29]、N_2O_4 和 HNO_2 等其他的活性氮(RNS)。N_2O_3 和 HNO_2 容易与生物分子上的氨基、巯基等亲核基团发生反应,生成亚硝基硫醇和亚硝胺等产物,亚硝基硫醇可以起到信号传递的作用[30],亚硝胺具有致癌性[31,32]。

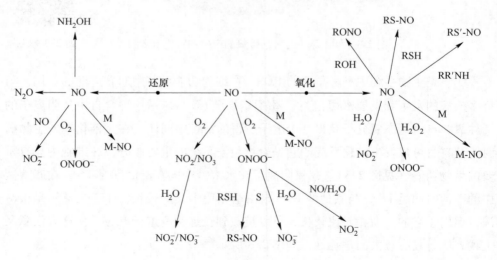

图 2.4 一氧化氮的氧化还原反应及其产物

一氧化氮在生物体内的产生,主要是通过一氧化氮合成酶(NOS)催化合成。在有氧和 DANPH 存在的情况下，NOS 能够促使 L-精氨酸转化生成 L-瓜氨酸和 NO[33,34]，其反应如图 2.5 所示。生成的 NO 会及时与细胞质内的鸟苷酸环化酶发生反应，增加环磷酸鸟嘌呤的生成量，因此发挥重要的生理功能。其次是通过黄嘌呤氧化酶[35,36]和硝基血管扩张剂[37,38]等产生 NO。

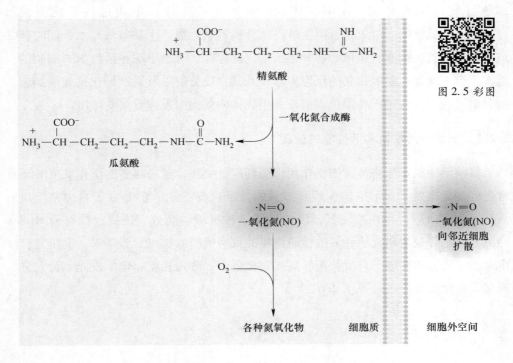

图 2.5 一氧化氮在生物体内的合成

一氧化氮也可以通过化学反应合成，例如用较低浓度的硝酸与铜反应来制备[39]。由于一氧化氮自由基非常活泼容易与氧气反应生成二氧化氮，因此收集一氧化氮时必须先排除氧气。通常是使用气体发生器产生一氧化氮自由基，利用氮气排除水溶液中的氧气后，再收集 NO 气体。制备一氧化氮的反应如下。

$$NaNO_2 + H_2SO_4 \Longrightarrow HNO_2 + NaHSO_4 \qquad (2.2)$$

其中 HNO_2 极不稳定，容易分解。

$$2HNO_2 \Longrightarrow NO\uparrow + NO_2 + H_2O \qquad (2.3)$$

（3）过氧化氢。过氧化氢（双氧水）的反应原理主要涉及其分解反应。在一般情况下，过氧化氢会自发分解成水和氧气：

$$2H_2O_2 \longrightarrow 2H_2O + O_2 \qquad (2.4)$$

这个反应是不可逆的，意味着一旦发生，过氧化氢就不能恢复到原来的状态。在漂白或其他应用中，过氧化氢的这种分解特性被用来达到预期的效果。然而，过氧化氢的分解速度相对较慢，为了加快反应速率，可以使用催化剂，如二氧化锰（MnO_2）、红砖粉末或氧化铜（CuO）。这些催化剂能够降低反应的活化能，从而加速过氧化氢的分解。在生物体内，过氧化氢的分解是通过过氧化氢酶（Catalase）催化的。过氧化氢酶能够迅速分解过氧化氢，生成氧气和水，同时释放出热量。这种酶在有机体的各种组织中广泛存在，有助于防止活性氧对细胞的损害。综上所述，过氧化氢的反应原理涉及其自发分解以及通过催化剂或酶加速的分解过程，这些过程在漂白、消毒和生物体内代谢中发挥着重要作用。

2.2.2 细胞中活性氧和活性氮的检测

目前活性氧对生物体的损伤作用已受到广泛关注，越来越多的研究发现 ROS 含量的变化有助于揭示机体病理状态下的分子机制[40,41]，但是由于自由基反应活性强、寿命短、存在浓度低，因此探寻一种灵敏、高效、便捷、准确的 ROS 检测方法已经成为研究活性氧的性质与相关致病机制的重点。近年来，测定氧自由基的主要方法有电子自旋共振、高效液相色谱、分光光度、化学发光、荧光分析法及电化学等方法（见图2.6）[42]。

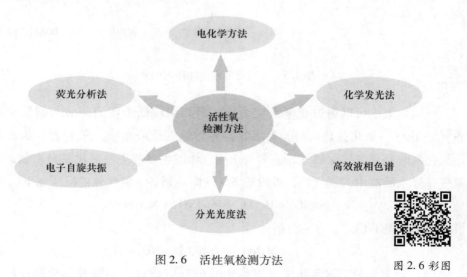

图 2.6 活性氧检测方法

图 2.6 彩图

（1）电子自旋共振法。电子自旋共振法（Electron Spin Resonance，ESR）或称电子顺磁共振（Electron Paramagnetic Resonance，EPR），是检测活性氧直接且

有效的方法[43],此方法主要检测一些含有未成对电子的自由基等顺磁性物质[44,45],其基本原理是利用捕获剂与活泼活性氧自由基发生反应形成较稳定的自由基(自旋加合物),然后再使用 ESR 进行检测。具有高灵敏度,测试样品不会受损及对化学反应没有干扰等优点。这一分析方法已在许多领域内越来越广泛应用。但由于活性氧自由基寿命极短,一般使用自旋捕获技术进行直接检测,可以作为检测自由基中间体和体内自由基检测的首选方法。Tada 等人报道了利用 ESR 定性和定量分析评估体内 ROS 的生成[46]。但是,电子顺磁共振检测仪器测试费用高,仪器价格高,被检测的样品需要低温冷藏,其软件测试结果需要专业人员分析,对于日常实验检测难以广泛应用[47]。

(2) 高效液相色谱法。高效液相色谱(High Performance Liquid Chromatography, HPLC)可以对生物样品中的自由基进行测定,其自身具有较好的分离能力,并且可以与质谱、电化学等检测方法一起应用。此方法仅适用于羟基自由基(·OH)。HPLC 是一种间接检测 ROS 的手段,但是由于这种方法仪器昂贵,检测操作程序繁琐。因此,该方法对于大范围推广存在一定的困难。

(3) 分光光度法。在采用 ESR、HPLC 等测试方法都存在缺陷的情况下,用分光光度法测定自由基被认为是比较简便实用的方法。其原理是利用 ROS 具有的氧化或还原特性,通过反应物发生氧化或还原反应,使生成物在紫外可见光谱吸收范围内测定其最大的吸收峰,进行间接测定样品中 ROS 的含量。经典的分光光度法是测定 $O_2^{·-}$ 含量,常选用苯三酚氧化法,在 420 nm 处测定其最大吸收波长。Armstrong 等人介绍了用分光光度法检测细胞线粒体内活性氧[42]。该方法操作简单,费用低,一般实验室都具有配备条件,但是检测的检测限、灵敏度、抗干扰能力是有待提高之处。

(4) 化学发光法。化学发光法(Chemiluminescence, CL)是分子发光光谱分析法中的一类,它主要是依据化学检测体系中待测物浓度与检测体系的化学发光强度在一定条件下呈线性定量关系的原理,利用仪器对体系化学发光强度的检测,进而确定待测物含量的一种痕量的分析方法。它可以利用发光试剂与活性氧反应后,根据发光的强度确定样品中 ROS 的含量。在有氧条件下,黄嘌呤/黄嘌呤氧化酶产生的 $O_2^{·-}$ 可与化学发光剂鲁米诺(Luminol)发生反应[48,49],对其产物进行激发,当返回基态时就会产生化学发光。与电子自旋共振和高效液相色谱法相比,发光法操作简便,设备低廉,检测快速。本方法对于研究自由基的产生,筛选清除自由基的药物,预防与治疗自由基有关的疾病具有一定的应用价值。这种方法具有的不足是它不仅可以检测 $O_2^{\cdot-}$，还可以检测·OH，$1O_2$ 和 H_2O_2，其首要难题是研究出专一性强的荧光素。

（5）荧光光度法。荧光光度法是一种灵敏的、可提供细胞内靶分子时空信息的 ROS 检测法。荧光探针一般都具有很强的专一性和很高的灵敏度，而且是一类无色、无荧光且稳定存在的染料分子，因此已被广泛应用。其原理是荧光探针与 ROS 反应后，探针的化学结构发生变化生成具有强荧光的物质，通过反应产物的荧光强度反映 ROS 水平，也可借助图像强度分析软件对荧光强度进行量化。目前，应用最广泛的荧光探针主要有 Dichlorofluorescin diacetate（DCFH-DA），3-Amino，4-aminomethyl-2′,7′-difluorescein，diacetate（DAF-FM DA），双氢罗丹（Dihydrorho-damine），氢化溴乙非锭（Hydroethidine，HE）等。

目前应用范围较广的活性氧荧光探针是 DCFH-DA。其原理是 DCFH-DA 本身没有荧光，可以自由地穿过细胞膜，进入细胞后，可以被细胞内的酯酶水解生成 DCFH，而 DCFH 不能穿过细胞膜，从而使探针装载到细胞内。细胞内的活性氧可以氧化无荧光的 DCFH 生成有荧光的 DCF。通过检测 DCF 的浓度就可以知道细胞内活性氧的浓度[50,51]。Gao 等人[52]利用 DCFH-DA 装载紫外光诱导的角质化细胞，间接评价细胞凋亡过程中 ROS 的产生水平。Li 等人[53]报道了在抗坏血酸协同 As_2O_3 诱导肝癌细胞凋亡实验中，使用流式细胞术和 DCFH-DA 荧光探针检测细胞内 ROS 活性氧水平的增加。

DAF-FM DA 荧光探针是最新一代用于一氧化氮定量检测的荧光探针。DAF-FM DA 可以穿过细胞膜，进入细胞后可以被细胞内的酯酶催化生成不能穿过细胞膜的 DAF-FM。DAF-FM 本身仅有很弱的荧光，但在与一氧化氮反应后可以产生强烈荧光，激发波长为 495 nm，发射波长为 515 nm。任何可以检测荧光的仪器，包括荧光显微镜、激光共聚焦显微镜、流式细胞仪、荧光分光光度计或荧光酶标仪都可以用该荧光探针检测。Chen 等人[54]利用 DAF-FM DA 荧光探针研究海洋微生物来源新型化合物 Xyloketal B 产生的一氧化氮。Liu 等人[55]研究花青素对圣经母细胞瘤 N2a 氧化应激的保护作用，对细胞进行预处理后用 DAF-FM DA 荧光探针装载细胞进而检测 DAF-FM 的荧光强度，间接验证细胞产生 NO。荧光光度法不需要对被测物质进行预处理，操作简单，费用低廉，但是这种技术不可避免也存在一些缺点，如灵敏度、选择性还有待进一步的提高。

（6）电化学法。体内活性氧的寿命短，浓度低，需要检测设备具备能够快速、灵敏、准确等性能，由于电化学方法能够实现活性氧浓度进行实时、直接、连续检测，已成为迄今为止唯一的检测活性氧的方法。电化学法为 ROS 的实时检测和生物组织样品分析提供了一种实用、简便的方法。该方法可以实现 ROS 的连续、准确及其活体组织的直接检测，而且不会损伤活体组织。Guo 等人[56]通过构建多功能层状石墨烯/过氧化物仿生酶/细胞外基质蛋白纳米结构生物界面，层状石墨烯的优良导电性以及其提供的立体界面用于细胞生长，仿生酶和细胞外基质使细胞具有良好的贴附和生长能力。此电化学传感器实现了对细胞释放的过氧化氢分子的原位检测。另外，Guo 等人[57]利用柔性的石墨烯薄膜为基体，通过与生物分子的复合，构建了一个具有生物相容性和特异催化性的功能柔性薄膜，实现了对一氧化氮的原位检测。Luo 等人[58]制备了一个基于 SOD 仿生酶与高导电 TiO_2 纳米针的直接电子转移的超氧阴离子电化学传感器，其对超氧阴离子检测具有稳定性好，选择性优异，实现了对正常细胞和癌细胞超氧阴离子释放的检测。电化学法终将成为生物医学工作者研究 ROS 引起的一系列疾病的有效手段。

早在 1992 年，Malinski 等人在自然杂志上报道了通过电化学催化氧化法对 NO 进行检测[59]，使用四（3-甲氧基-4-羟基苯基）卟啉的 Nafion 膜氧化电聚合的方法修饰到碳纤维电极表面，对 NO 具有很好的选择性，测得 NO 的检测限为 10 μmol/L。随后，Ohsaka 等人成功实现了 $O_2^{\cdot -}/O_2$ 电对的氧化还原反应，用于水溶液中 $O_2^{\cdot -}$ 的检测[60]。在一定的条件下，氧化还原电对 $O_2^{\cdot -}/O_2$ 的电化学行为表现出良好的可逆性，进而开展了 $O_2^{\cdot -}$ 的电化学传感器[58,61-63]。电化学技术已被证明是用于体内分析物量化的有效方法[64]，其与常规的方法相比具有如下几个优点：第一，对于 ROS 的电化学测量，不需要其他的化合物指示剂；第二，电化学技术可以对短寿命的 ROS 即时响应；第三，电化学提供了一种原位实时监测 ROS 的途径。因此，电化学生物传感器用于体内活性氧的检测具有潜在的应用前景，有望取代传统的方法。

综上所述，自由基的测定一般是利用某些体系产生自由基或本身就是自由基，再通过某些反应体系产生的颜色变化、发光现象、电化学行为等间接方法测定自由基的含量。迄今为止，化学反应法和捕获法的分析选择性以及专一性还难以令人满意，能够直接进行测定自由基的方法唯有电化学方法，其在检测自由基方面具有高灵敏度、高选择性，而且可以实现生物活性内的直接实时、原位检测。在生物医学和环境科学等领域，特别是自由基在生物活体内的直接原位测定，对于解释生命机体的各种疾病病理有着重要的意义，这也是诸多学者研究的目标。

2.2.3 细胞中活性氧和活性氮的清除

自由基是人体生命活动中的中间代谢产物，具有高度的化学活性，是机体有效的防御系统，如果不能维持在稳定的水平会影响机体的生命活动。但是活性氧产生过多不及时清除，它就会攻击机体内的生物分子与各种细胞器，导致机体氧化损伤，诱发各种疾病。活性氧自由基清除剂（FRS）是指具有延迟、抑制和阻断 ROS/OFR（氧自由基）氧化损伤的物质的总称，能够与 OFR 结合并使之清除的机体保护剂。人的生命活动离不开自由基，但是如果体内自由基过多或清除过慢，自由基就会攻击并损坏大分子，对细胞膜、核酸及机体蛋白质等造成损伤，这是引起机体衰老的根本原因，也是诱发恶性肿瘤等许多疾病的重要起因。在机体正常运行过程中保护细胞和组织免受氧化损伤具有重要作用。生物在进行新陈代谢过程中是离不开氧分子的，可是当活性氧浓度很低时对机体也是有害的，为了更好地生存，生物体在长期的进化过程中形成了防御活性氧自由基侵害的抗氧化系统，在此过程中主要是由氧化酶和抗氧化酶参与其中并发挥重要调控作用（见图 2.7），以维持机体的氧化还原平衡的。这些抗氧化系统是通过产生一些抗氧化剂来抵御 ROS 对机

图 2.7 彩图

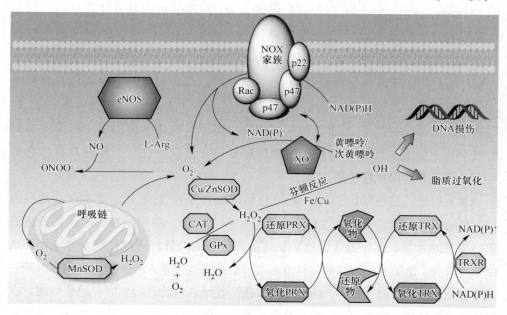

图 2.7 机体中的氧化酶和抗氧化酶[67]

体的伤害。其防御机理是通过清除或抑制机体产生的 ROS，进而阻止 ROS 参与机体的氧化损伤过程[65,66]。

通常将自由基触发的氧化应激清除剂分为酶类清除剂和非酶类清除剂[68]（见图2.8）。酶类清除剂主要有超氧化物歧化酶（SOD）、过氧化氢酶（CAT）

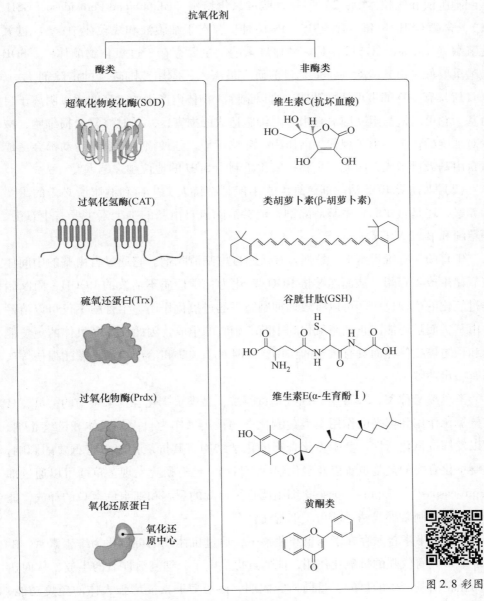

图 2.8　抗氧化剂（自由基清除剂）的分类

和谷胱甘肽过氧化物酶等。非酶类清除剂主要有生育酚、圣草酚、血红蛋白、维生素 C、β-胡萝卜素、微量元素硒等。具体介绍如下。

（1）酶类 ROS 清除剂。在正常机体中存在着多种能够抑制或清除 ROS 的酶，主要有：1）超氧化物歧化酶（SOD），有 Cu、Zn-SOD 和 Mn-SOD 三种，分别存在于细胞胞液和线粒体中；2）谷胱甘肽过氧化物酶（Glutathione Peroxidase，GSH-Px），含硒 GSH-Px 和不含硒 GSH-Px 两种，存在于细胞液和线粒体中；3）过氧化氢酶（Catalase，CAT），是一种血红素酶，主要存在于过氧化物酶体中。SOD 是最重要的自由基清除剂，也是目前研究最深入，应用范围较广的清除剂之一。SOD 所具有特殊的生理活性功能，可以清除生物体内产生的过多的超氧阴离子自由基。它可对抗与阻断因氧自由基对细胞造成的损害，并及时修复受损细胞，减少对细胞的伤害。由于现代生活压力，环境污染，各种辐射和超量运动都会造成氧自由基大量形成，因此，生物抗氧化机制中 SOD 的地位越来越重要。

（2）非酶类 ROS 清除剂。维生素不但是维持人类生命和健康所必需的重要营养素，还是重要的自由基清除剂，能够清除氧自由基的维生素主要有生育酚、圣草酚和 β-胡萝卜素。

生育酚又称维生素 E，是强力有效的自由基清除剂。它通过自由基的中间体被氧化生成生育醌，从而能够将 ROO·转化为化学性质不活泼的 ROOH，阻断脂类过氧化的连锁反应，可以有效地抑制脂类的过氧化作用。生育酚不仅可以清除自由基，而且还能防止油脂氧化和中断亚硝胺的生成，因此在提高机体的免疫能力和预防癌症等方面具有重要的作用，同时对于预防和治疗缺血再灌注损伤等疾病有一定功效。

圣草酚又称3′，5，4′，7-四羟基黄酮烷，是类黄酮化合物家族中的一员，具有抗氧化作用和抗炎的作用。类黄酮化合物被用来作为自由基受体和链终止剂，可以发挥抗氧化活性，黄酮类化合物的化学结构对其抗氧化能力有重要的影响，主要指化合物的羟基位置以及羟化程度。目前，已有研究发现圣草酚可以通过抑制 pro-caspase-3 或 pro-caspase-9 细胞凋亡蛋白酶的裂解和细胞色素 C 的释放，能够有效地抑制紫外线诱导的角质细胞死亡[69]。

β-胡萝卜素普遍存在于水果和蔬菜中，通过机体代谢可转化为维生素 A。β-胡萝卜素具有较强的抗氧化作用，能通过提供电子，抑制活性氧的生成，从而达到防止自由基产生的目的。目前研究表明，β-胡萝卜素能增强人体的免疫功能，防止吞噬细胞发生自动氧化，增强巨噬细胞、细胞毒性 T 细胞、天然杀伤细胞对肿瘤细胞的杀灭能力。在多种食品中，β-胡萝卜素与不饱和脂肪酸的稳定性密切相关。

2.2.4 细胞中活性氧和活性氮对机体的危害

活性氧与机体细胞的许多生理功能活动以及各种疾病的发生密切相关，正常情况下体内的氧化还原处于动态平衡过程中对抗炎和肿瘤的抑制等具有重要意义。细胞在正常新陈代谢过程中或接受高能辐射、药物和化学污染物等作用后都可以产生活性氧。在外源刺激下，一旦机体自由基过量产生就会引起细胞质膜、蛋白质和DNA的氧化损伤[70,71]，导致机体的代谢平衡紊乱或者氧化防御平衡失调，RONS就会促使生物体内的生物分子引起氧化损伤。依据分子生物学理论，生物分子主要是由于脂质过氧化而遭受损伤，主要包括有生物膜、蛋白质和DNA的损伤。因而发生一系列疾病，如癌症、动脉硬化、糖尿病、心脑血栓、动脉粥样硬化、高血压等疾病[72-75]，引起人们对活性氧的普遍兴趣，从而激发了人们更深入地去研究活性氧的各种特性，开发各种相关的抗肿瘤药物、抗氧化、抗衰老物质，在医学和分子生物学领域已成为一项广泛引起重视的研究课题。总体概括起来阐述可以归纳为三个方面。

（1）RONS对生物膜的损伤：生物膜主要由蛋白质、脂质和糖类组成。生物膜具有以下作用：第一，可以使细胞具有一个相对稳定的内部环境，同时在细胞与外部环境的物质运输、能量转换和信号传递中起着决定性作用。第二，许多重要的化学反应都是在生物膜上进行的，由于化学反应需要酶的参与，广阔的膜面积为多种酶提供了大量的附着点。第三，细胞内的生物膜可以把各种细胞器分隔开，使细胞内多种化学反应能够同时进行而且互不影响，保证了细胞生命活动的高效和有序进行。生物膜中含有多种多不饱和脂肪酸（Ployunsaturated Fatty Acid，PUFA），它是最容易受RONS攻击的生物分子，易发生过氧化作用，引起自由酸的减少。在正常情况下，PUFA是液态的流动性较好，而当RONS攻击时使膜的流动性降低、膜的通透性增加，严重时甚至膜完全裂解，以至于细胞稳定的内环境被打破。此外，生物膜所含的PUFA过氧化会使细胞变得僵硬，细胞膜失去弹性和变形性，导致生物膜的功能异常。

（2）RONS对蛋白质的损伤：蛋白质组分是生物体细胞的重要组成成分，其本身或组成蛋白质的氨基酸均是RONS攻击的靶分子。RONS对蛋白质的损伤主要表现在肽链的断裂、蛋白质的交联、使蛋白质的结构和活性位点变化。蛋白质的交联主要是指使蛋白质分子发生分子内或分子间交联，而蛋白质的结构破坏是指使双螺旋结构中的折叠减少，没有规律的折叠增加。当蛋白质受到这些变化后，必然会引起生物功能的巨大变化。此外，蛋白质也会与糖类发生交联使酶活性消失，引起细胞膜结构的变形性下降，引起细胞的衰老与死亡。

（3）RONS对核酸的损伤：许多核苷酸聚合而成的生物大分子化合物称为核酸，其是生命的重要遗传物质。脱氧核糖核酸（DNA）在生物学中具有重要的作用，它是遗传信息的载体，起着储存、复制和传递遗传信息的重要作用。RONS对DNA的损伤主要表现在以下几个方面：1）碱基的修饰。2）氢键的断裂。3）DNA自身交联或者DNA与蛋白质的交联。这是由于RONS诱导的脂质过氧化产物较易与DNA碱基进行共价结合，生成稳定的氧化产物。这些因素导致DNA的结构或序列发生变化，使其携带的遗传信息发生变化，对DNA的功能与遗传性引起很大的影响，最终发生基因突变或者引起细胞的癌变，导致相关疾病的发生。

在生物体内外，RONS不仅可以激活信号传导通路，还通过对氧化还原敏感基因的诱导活化影响细胞的生理功能，因此，RONS在诱导细胞凋亡的过程中起着关键作用，同时对于正常细胞生理功能的稳定也需要ROS参与对细胞的增殖、分化和凋亡的调控。具体是由于RONS介导的氧化应激能够扰乱生物体内自由基的生成与清除，打破氧化还原平衡状态，最终导致机体受到氧化损伤。高浓度的活性氧能够引起生物膜、蛋白质和核酸的氧化损伤，对机体的免疫系统和结构功能产生影响。氧化与抗氧化酶处于动态平衡状态，正常情况下细胞代谢需要RONS，如低浓度的RONS能促进细胞增殖[76-78]。但是当高剂量的RONS刺激细胞时，导致机体平衡失调使RONS产生增多，细胞内自身的抗氧化系统无法抵抗这种高强度的氧化损伤，细胞就会发生一系列的生理变化，导致细胞器功能失调，如线粒体的膜孔被打开使通透性增加，蛋白质随即释放出来，即可导致细胞凋亡[79,80]，甚至使细胞损伤引起细胞病变[81,82]。

目前，大量研究发现活性氧诱导的细胞凋亡是多种凋亡信号相互联系、紧密结合与共同作用的结果，但是具体的调节机制目前尚不清楚，研究的最深入、最清楚的凋亡信号通路主要是半胱天冬酶（Caspase）依赖性细胞凋亡途径，包括线粒体通路和死亡受体介导的信号通路，这两种信号通路简要介绍如下。（1）线粒体途径是一种内在的凋亡途径[84]。线粒体介导的细胞凋亡信号分子是一种主要的凋亡途径，由于线粒体是RONS生成的主要部位及其作用的主要靶点，二者之间有密切的关系，因此，外源性 RONS 经过线粒体信号通路介导细胞凋亡的发生。当 ROS 刺激细胞时可增加 JNK，p38 和 ERK1/2 激酶活性的磷酸化，并通过调节细胞质中的 BCL-2 家族蛋白的表达水平[85,86]和线粒体膜的去极化，使线粒体膜通透性改变，膜孔打开，位于线粒体膜间隙的蛋白如细胞色素 C[87]，凋亡诱导因子（AIF）通过膜孔释放到胞质中，诱导 Caspase 级联反应[88]和胱冬肽酶-9（Caspase-9）的酶活性被激活，引起细胞凋亡。（2）死亡受体介导的细胞凋亡信号通路是一种外在的凋亡途径。它是由死亡受体 Fas、TNF 受体途径，经相应的配体 FasL、TNF-α 诱导激活，引发 Caspase-8 与 Caspase-10 的激活，诱导细胞发生凋亡[88]。综上所述，ROS 诱导的细胞凋亡与上述两条信号通路是密切相关的，具体分析可从几个方面一一论证：（1）活性氧是生物在经受体内外界环境相互作用的条件下产生的，如药物、紫外线辐射、环境污染、内分泌紊乱等，通常可以选用一些凋亡抑制剂来消除活性氧对细胞的损伤；（2）在体外检测实验中通过加入一定量的 ROS（主要指 $O_2^{\cdot-}$、NO 与 H_2O_2）或对细胞内抗氧化剂进行消除可以诱导细胞凋亡；（3）ROS 清除剂可以用于抑制或减缓 ROS 诱导的细胞凋亡[89]。

2.2.5 细胞活性氧和活性氮在抗肿瘤药物研究中的应用

RONS 作为物种在生命系统中可能有害或有益的双重角色，其在人体的病理学与生理学中发挥着极其重要的作用[67]。过量的活性氧自由基可以引起细胞氧化损伤，其结果对生物分子引起严重的损伤，包括脂类、蛋白质和 DNA[40,90]。此外，它可以破坏细胞内的氧化还原平衡，降低线粒体膜电位，引起细胞的死亡和凋亡[91-95]。通过内源性或外源性来源不受控制的 RONS 形成被称为氧化应激，是许多类型癌细胞的共同特征。氧化应激标志物的增加证明了癌细胞中氧化应激的发生，而癌细胞中 RONS 水平的升高可以通过抗氧化防御机制来抵消。几十年来，科学家们一直认为正常细胞向癌细胞的转化及其存活取决于癌基因的激活或肿瘤抑制基因的失活：这一假设被称为"致癌成瘾"[96]。因此，基于针对致癌基因和/或肿瘤抑制基因的药物开发了药理学干预措施。然而，最近的研究表明，靶向维持癌细胞存活的重要机制是一种很有希望的癌症治疗方法[76]。由于癌细胞能够适应其快速生长所导致的氧化还原稳态的改变，并且可以发展替代代谢途径，使其对外源性应激源（包括化疗和放疗[97]）不那么敏感，因此通过靶向治疗有效调节氧化应激可能为癌症治疗提供了一个有希望的策略[98]。

生物系统中活性氧的产生如图 2.9 所示。

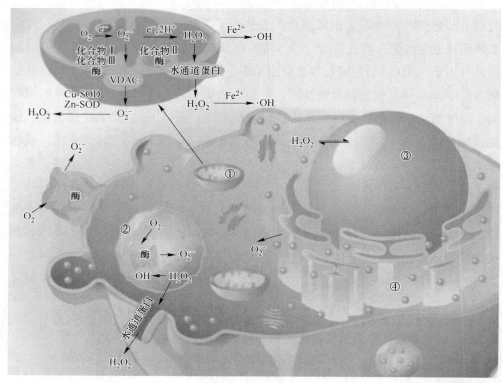

图 2.9　生物系统中活性氧的产生

①—线粒体；②—细胞溶质；③—细胞核；④—内质网

图 2.9 彩图

2.3　基于纳米酶的电化学生物传感器在活性氧检测中的应用

目前适用于生物体 $O_2^{\cdot-}$ 检测的电化学传感器主要包括两类：第一类是酶（超氧化物歧化酶或细胞色素 C）标记的电化学生物传感器；第二类是由能够催化 $O_2^{\cdot-}$ 还原的有机和无机纳米材料组成的纳米酶电化学传感器。在此，主要综述第二类纳米酶电化学生物传感器在 $O_2^{\cdot-}$ 和 H_2O_2 检测中的应用研究。从生物学的角度来看，天然酶的活性位点被定义为底物分子发生特定化学反应并显著降低活化能的区域。它是负责酶功能的关键成分。因此，获得与天然酶相当的催化性能的最有效方法是模拟它们的活性位点。由于纳米材料所表现出的独特的物理化学特性，包括它们的微小尺寸，膨胀的表面积和显著的反应性。迄今为止，各种纳米材料在构建电化学纳米酶用于检测 ROS 方面的应用越来越广泛。

2.3.1 基于过渡金属的纳米酶电化学生物传感器

尽管基于酶的传感器具有优势，但它们的缺点促使研究人员寻找其他类型的具有更好稳定性、高选择性和低成本的改性剂。一个很好的替代方案是基于所谓的"仿生酶"的传感器，这是非蛋白质成分，能够表现出与 SOD 相当的对 $O_2^{·-}$ 的催化活性。这类材料包括基于过渡金属或贵金属及其纳米材料的材料，以及金属有机框架（MOF）。纳米碳材料，包括碳纳米管、纳米颗粒、石墨烯等具有高强度、优异导电性、高电流密度下高效的电荷转移能力、低电阻、大表面积和卓越的化学稳定性等特点，这些固有的特性使它们特别适合用于纳米酶标记的电化学生物传感器。

基于过渡金属的纳米酶传感器。锰是一种过渡金属，对 $O_2^{·-}$ 毒性具有非酶保护作用。锰基传感器的工作原理是 $O_2^{·-}$ 还原为 O_2 和 H_2O_2 被 Mn^{2+} 离子吸收。磷酸锰（$Mn_3(PO_4)_2$）因其生物相容性而在电化学传感器领域获得了相当大的关注，适合生物医学应用[99]。此外，由于其独特的催化和电子特性，$Mn_3(PO_4)_2$ 在材料科学中具有举足轻重的重要性。然而，基于过渡金属磷酸盐纳米酶的Ⅱ电化学传感器的性能受到材料的微观结构和尺寸的显著影响。为了应对这一挑战，已经探索了各种策略。Wang[100]利用微乳液法合成了具有可调孔结构的纳米结构 $Mn_3(PO_4)_2$ 空心球，所得纳米结构 $Mn_3(PO_4)_2$ 空心球具有高表面积和多孔结构，对 $O_2^{·-}$ 的检测具有优异的催化性能。在另一项研究中，Ding 等人[101]开发了石墨烯/DNA/$Mn_3(PO_4)_2$ 仿生酶。石墨烯作为载体被证明对增强 $Mn_3(PO_4)_2$ 纳米颗粒的催化活性非常有益。Wang 等人[102]开发了一种基于 $Mn_3(PO_4)_2$ 纳米颗粒与壳聚糖结合的传感器，选择壳聚糖可以定制传感器材料的纹理并提高其生物相容性。将得到的微球滴在丝网印刷电极（SPE）上，然后在电极表面培养 4T1（小鼠乳腺肿瘤细胞系），可以原位检测活细胞释放的超氧阴离子。与其他活性氧（H_2O_2）、离子（Cl^-、K^+、SO_4^{2-} 和 Na^+）和生物分子（抗坏血酸、尿酸和葡萄糖）相比，所开发的传感器对 $O_2^{·-}$ 具有较高的选择性。使用这种方法，作者获得了相当低的检测限（9.40×10^{-9} mol/dm³）。总体而言，锰基仿生传感器对

$O_2^{·-}$ 的选择性高[103]，保质期长（约30天），易于修饰[104]，成本低，且具有纳摩尔水平的低检测限。锰基传感器的主要缺点是改性剂的合成过程复杂且耗时长。Olean-Oliveira 等人报道了一种非酶化学电阻器的发展用于检测超氧自由基的传感器，利用偶氮聚合物结合还原氧化石墨烯（rGO）作为传感器的电阻平台应用[105]。该传感器平台采用聚偶氮-有序沉积的方法制备俾斯麦棕Y（azo-BBY）和还原氧化石墨烯薄膜使用逐层组装技术（见图2.10(a)）。所得到的纳米复合膜表现出有趣的协同性能，结合了偶氮聚合物的氧化还原性能和石墨烯优异的电子导电性和稳定性。实时阻抗测量（时阻抗）采用 poly(azo-BBY)/rGO 传感器呈现线性关系真实阻抗与超氧阴离子浓度之间（范围从 0.12~2.6 mmol/L），检出限为 81.0 μmol/L。Xuan 等人进一步介绍了一种制造模拟酶的无金属催化剂的新方法，专门用于电化学检测 $O_2^{·-}$，将磷酸基团掺入石墨烯泡沫中[106]（见图2.10(b)），通过无模板水热反应进行合成，涉及用不同数量的植酸（PA）处理氧化石墨烯（GO）得到三维多孔石墨烯基泡沫（PAGF）。该实验结果证实了传感器研制成功，并得到了有效的应用。用于测定细胞释放的氧气，表现出卓越的细胞氧动态监测性能。介孔碳材料具有大量的边缘平面状缺陷位点，这有效地促进了电子向被分析物的转移，提高了电极界面的电化学活性。中空介孔碳球（HMCSs）的碳壳内介孔通道的存在为传感器和分析物之间的质量传递和/或电荷转移提供了有利的特性。研究采用掺氮中空介孔碳球（N-HMCSs）修饰丝网印刷碳电极（SPCE）开发了一种无酶无金属的电化学传感器，其在检测 $O_2^{·-}$ 过程中展现出良好的灵敏度[107]。通过计时电流响应曲线记录了 N-HMCSs/SPCE 在 $-0.15\ V$ 下对 $O_2^{·-}$ 的电流响应（见图2.10(c)）。随着 $O_2^{·-}$ 的加入，电流表现出不同的变化，并与 $O_2^{·-}$ 浓度成正比关系，具有较宽的检测范围，最高浓度可达到 480 mmol/L。基于电极的工作面积（0.071 cm^2），计算出 N-HMCSs/SPCE 的灵敏度为 1493.2 mA/(cm^2·mmol·L^{-1})，对 $O_2^{·-}$ 的检测限为 2.2 μmol/L。

H_2O_2 检测。气态过氧化氢作为一种生物标志物，与肺癌和哮喘等严重疾病有关，发挥着重要作用。监测呼出气体生物标记物的常规方法包括利用呼出气体冷凝物以及标准分析技术。在这种情况下，Klun 等人[108]提出了一种新的 H_2O_2 气体传感方法。将含有 Cu(Ⅱ) 离子的聚丙烯酸酯凝胶电解质作为传感材料，促进了气态分析物的积累和稳定。通过与 Cu(Ⅱ) 离子的氧化还原作用，快速、灵敏地检测出 H_2O_2。值得注意的是，Klun 等人开发的气体传感器在环境条件下仅需 2 min 的积累即可成功且快速地检测到气态 H_2O_2[108]。该传感器在较低的浓度

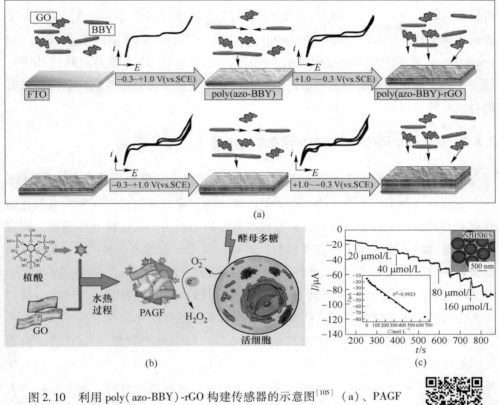

图 2.10 利用 poly(azo-BBY)-rGO 构建传感器的示意图[105]（a）、PAGF 材料的合成及其在细胞释放超氧阴离子检测中的应用[106]（b）和在 −0.15 V 下连续向 0.1 mol/L 除氧的 PBS（pH=7.4）中注入 $O_2^{·-}$ 时 N-HMCSs/SPCE 的电流-时间响应（c）[107]

图 2.10 彩图

范围内表现出良好的灵敏度，并且在 10~100 mg/m³ 的浓度范围内表现出广泛的线性响应。这一研究突破不仅推动了 H_2O_2 检测领域的发展，而且拓展了在爆炸物检测、环境监测、职业健康安全等新兴领域的潜在应用。通过提供一种快速、灵敏、可靠的 H_2O_2 气体传感方法，这项工作为各种类型的 H_2O_2 气体传感打开了大门。Rao 等人[109]在气液界面采用改良银镜反应制备了 Ag/FePO₄ 纳米酶。FePO₄ 纳米球具有高的表面体积比和表面负电荷，使其具有更大的表面积来装载更多的 Ag 纳米粒子（NPs），并增强了还原 H_2O_2 的催化性能。电化学传感器表现出良好的分析性能，具有线性检测范围宽（$3.0\times10^{-5} \sim 1.1\times10^{-2}$ mol/L）和较低的检测限。该传感器还具有可接受的再现性和抗干扰能力。Peng 等人[110]通过钴与钴之间的直接反应原位合成 $Co_3(PO_4)_2 \cdot 8H_2O$ 纳米酶Ⅱ和磷酸盐。研究人员证实，$Co_3(PO_4)_2 \cdot 8H_2O$ 纳米酶具有过氧化物酶样活性。研究进一步揭示了过氧化物酶样活性服从 Michaelis-menten 动力学，Michaelis 常数（K_m）为 0.073 mmol/L；表明与天然辣根过氧化物酶相比，它对 H_2O_2 的亲和力更高（HRP）。综上可知，新型纳米酶可方便地就地合成，从而消除了苛刻的要求，费力、耗时的合成和纯化过程。

2.3.2 基于贵金属的纳米酶电化学生物传感器

在过去的五年中，以金属及其化合物为主要成分的材料作为催化 $O_2^{\cdot-}$ 还原的主要成分得到了广泛的应用。对类似改性剂的兴趣增加与它们强大的电催化活性有关，从而产生具有高稳定性和低成本特征的超敏感和高选择性传感器。除锰外，其他过渡金属也用于制造改性剂，例如，基于钴的纳米材料[111-113]或氮掺杂石墨烯中分散的钴原子[114]。此外，在铁化合物的应用中获得了有趣的数据 rGO-非晶碳纳米纤维[115]贵金属是具有仿生特性的有吸引力的材料酶。与过渡金属相反，它们表现出化学惰性，这使得它们适合在不同的介质中使用。

贵金属纳米材料，如金纳米粒子（AuNPs）和银纳米颗粒（AgNPs），由于它们的特性，稳定性显著，电导率高，制备简便，比表面积大，以及出色的生物相容性引起了科学界的极大兴趣。值得注意的是，新兴研究强调了纳米颗粒的大小、形状和分布对其电催化活性的重要影响。这些发现强调了调整这些参数以优化贵金属纳米材料电化学性能的重要性。铂[116]或铂和银[117]纳米颗粒被用作其他贵金属材料来制造仿生酶。橙皮苷-铜络合物也被用作酶模拟物用于检测巨噬细胞活性[118]。基于金属及其纳米材料的传感器具有高度选择性和生物相容性，与酶传感器相比，具有惊人的长期稳定性（长达 120 天）[114]，并且可以达到超低氧检测限，低至 10^{-15} mol/dm^3[119]。

检测 $O_2^{\cdot-}$。Fan 等人通过热解一种新型银基金属有机骨架，成功获得了均匀分散的银纳米颗粒（AgNPs），他们的研究报道如下[120]。具体而言，以银为金属中心，以苯并咪唑为有机配体，对银基金属-有机骨架进行热处理，合成了 AgNPs@C 纳米复合材料。在获得的 AgNPs@C 纳米复合材料中观察到对 $O_2^{\cdot-}$ 还原的电化学响应，表现出超宽的线性范围（$3.032 \times 10^{-13} \sim 5.719 \times 10^{-5}$ mol/L）和极低的检测限（1.011×10^{-13} mol/L）。

H_2O_2 检测。Zhu 等人[121]从自组装肽纳米纤维（PNFs）中获得灵感，成功设计并合成了一种新型杂化材料 PtNWs-PNFs/GO，该材料由仿生石墨烯负载的超细铂纳米线（PtNWs）与 PNFs 集成组成。可控的自组装过程使 PNF 成为氧化石墨烯纳米片和 PtNW 之间的桥梁。PtNWs 以其高催化活性而闻名，它的存在赋予了 PtNWs-PNFs/GO 混合传感器显著的电化学性能。该传感器具有 0.05～15 mmol/L 的扩展线性检测范围和 0.0206 mmol/L 的低检测限。Ko 等人[122]将金属表面固定的 Au 和 Pt 双金属纳米颗粒琼脂糖微珠通过化学手段，得到一种称为杂化的纳米结构 Au@PtNP/GO。双金属纳米颗粒特别是氧化石墨烯赋予了混合纳米结构强烈的过氧化物酶催化活性，在 H_2O_2 存在下加入 3，3′，5，5′-四甲基联苯胺（TMB）进行反应。这种混合纳米结构实现了比色和电化学的双重应用检测。引入含 H_2O_2 的 TMB 底物溶液后发生 TMB 的催化氧化。因此，在电极表面，氧化 TMB 随后进行电化学还原，从而扩大了 H_2O_2 的检测范围，从 1 μmol/L 到 3 mmol/L 降低检测下限，量化为 1.62 μmol/L。此外，发达国家护理点（POC）设备表现出 H_2O_2 的准确测定，证明在使用人工尿液的真实样品测试中具有很强的重复性和再现性。

此外，构建贵金属纳米酶标记的电化学生物传感器通常是快速和简单的，其主要缺点是改性剂的合成过程复杂且耗时，以及一些贵金属的成本较高。

2.3.3 基于金属有机骨架的纳米酶电化学生物传感器

金属有机骨架（Metal-Organic Frameworks，MOF）由于其独特的性质，如高比表面积、可调节的孔隙结构和暴露的活性，引起了人们的广泛关注。MOF 由金属节点和有机配体组成，在电化学、荧光、比色、光电化学和电化学发光传感等领域有着广泛的应用。它们的特殊特性使它们非常适合这些应用，从而能够开发先进的传感平台和设备。MOF 的高比表面积有助于高效吸附分析物，并进行相互作用，而可调的孔结构允许选择性捕获和识别目标分子。此外，MOF 的暴露活性有助于其在各种传感模式下的性能增强。因此，MOF 作为下一代传感技术发展的有前途的材料显示出巨大的潜力。

检测 $O_2^{\cdot -}$。Zhang 等人[123]进行了一项研究，旨在通过开发一种简单的一步合成锰有机骨架（Mn-MOF）的策略来增强对 $O_2^{\cdot -}$ 的传感性能。有趣的是，他们通过仔细调整溶剂比例和调节初始前驱体浓度，实现了 Mn-MOF 纳米颗粒、不对称纳米棒棒糖和具有均匀成分的纳米棒的合成。随后，作者发现 Mn-MOF 纳米棒棒糖由于其更大的活性表面积，与其他纳米结构相比，表现出更优越的 $O_2^{\cdot -}$ 传感能力，这可归因于不对称结构提供的优异分散性，以及茎结构促进的加速电子转移速率。利用 Mn-MOF 纳米棒棒糖进行 $O_2^{·-}$ 检测，获得了 105 $\mu A/cm^2$ 的高灵敏度，成功实现了活细胞释放 $O_2^{·-}$ 的实时原位检测。该研究不仅为其他 MOF 纳米材料的溶剂工程形貌提供了有价值的见解，而且还促进了对 Mn-MOF 纳米结构在传感技术中的理解和潜在应用。

检测 H_2O_2。基于 MOF 的电化学传感器在检测 H_2O_2 方面得到了广泛的关注。Gao 等人合成了一种具有多酶活性的 Pt 纳米粒子修饰的金属卟啉 MOF（Pt@PMOF(Fe)），并利用其构建了电化学 H_2O_2 传感器[124]。Wei 等人的另一项研究合成了用于检测 H_2O_2 的 3D 基于 Co 的沸石咪唑酸框架（3D ZIF-67）[125]。Liu 等人报道了一种基于 Ti-mesh 的镍金属-有机框架纳米片阵列（Ni-MOF/TM）作为测定 H_2O_2 的无酶电化学传感平台的开发[126]。Yang 等人选择 MIL-47(V) 作为电催化剂，探索电化学感应 H_2O_2 的可行性[127]。然而，这些传统方法是在单输出模式下执行的，这使得它们容易受到外部干扰（如生物环境的复杂性、非标准化测试协议、操作人员或测试环境之间的差异）引起的假阳性或假阴性结果的影响。这些限制对单输出分析方法的准确性提出了实质性的挑战，并对其在疾病诊断中的实际应用施加了限制。

2.4 电化学生物传感器在活性氮检测中的应用

一氧化氮（NO）是一种非常具有化学活性的物质，可以被氧气或超氧化物 NO_2^- 或 NO_3^- 离子快速氧化[128]。由于 NO 具有扩散快、浓度低、半衰期短的特点，因此有必要开发一种高灵敏度快速输出的新方法来实时准确监测活细胞中 NO 的释放情况。NO 是细胞中另一种重要的信使分子，调节多种生物过程，如神经元细胞间通讯、血管舒张、抗炎血管松弛和免疫功能[129]。它可以与各种细胞内/细胞外靶点发生反应，修饰蛋白质（氨基酸）、DNA、脂质，并调节血管中的转录因子。所产生的 ROS 和 RNS 还可以改变各种靶激酶/磷酸酶的功能。NO_2^- 单电子还原为 NO 有几种酶促和非酶促途径。这些包括脱氧血红蛋白，脱氧肌红蛋白，黄嘌呤氧化酶和线粒体呼吸链的酶[130]。

纳米技术和纳米科学的快速发展，为电化学 ROS/RNS 传感器的纳米材料制造提供了许多新的创新方法。由于纳米材料具有体积小、表面积大、反应活性高等基本物理化学性质[131]，如贵金属基材料、过渡金属基材料、碳基材料等，越来越多的纳米材料被应用于构建电化学 RNS 传感器。近年来,由于细胞外 RNS 在细胞信号病理学中的重要作用,许多研究都关注于细胞外 RNS 的检测或定量。Ach,神经递质启动信号级联反应,激活钙调蛋白复合物,从而触发 NO 的生成和释放。

玻璃碳电极(GCE)是一种经典的刚性电极,由于其导电性好、硬度高、光洁度高、化学稳定性好等优点,通常被用于构造传感器。Kim 等人以 GCE 作为基本传感接口,通过聚合物-氧化石墨烯/氧化锌修饰,制造了一种传感器,用于检测多糖刺激 PC12 细胞释放的 NO[132]。然而,GCE 作为检测 NO 的传感接口,由于 GCE 与 NO 之间的距离较长,在扩散过程中受到损耗的限制。为了减小 NO 与传感界面之间的距离,重要的方法是在传感界面中培养活细胞或与传感材料共培养,但刚性电极由于其刚性限制了其使用。Xu 等人利用同样为刚性电极的氧化铟锡(ITO)电极构建了活细胞释放 NO 的传感器,灵敏度高达 0.21 μA/($\mu mol \cdot L^{-1} \cdot cm^2$),检测限低至 180 nmol/L[133]。为了减少 NO 与传感界面之间的距离,他们制备了 6 mm 厚的 PDMS,其中包含一个孔,用于建立临时腔室来培养细胞并促进细胞附着在刚性传感界面上。当大部分电池黏附在界面上后,稍微移除 PDMS 腔体进行电化学测量。这种方法为在刚性传感界面中培养细胞并立即捕获目标分子提供了一种新方法。Liu 等人合成了具有大宽高比的金纳米管(NTs)来构建可拉伸的电化学传感器[134]。该传感器具有良好的稳定性、电化学性能和生物相容性。研究结果表明,5 nmol/L NO 诱发的电流明显增加,计算出的检出限为 3 nmol/L(S/N = 3);展示了在无拉伸和拉伸状态下出色的电化学传感能力和量化 NO。Zhao 等人介绍了一种简单环保的紫外线(UV)照射辅助技术,构建可拉伸的纳米结构金膜作为检测 NO 的柔性电极[135]。柔性电极可以在 10 nmol/L ~ 1.295 μmol/L 的宽线性范围内检测 NO。此外,它记录了人脐静脉内皮细胞(HUVECs)在正常和机械应变状态下的 NO 释放。因此,由于这些类型的柔性传感器具有出色的机械顺应性,因此具有在体内监测目标分子的潜力。此外,由于这些经典电极的刚性和较大的弹性模量,这些传感器在细胞和组织变形时不具有机械顺应性,因此无法检测和量化机械变形的生化信号。这一现象限制了它们在体内定量中的应用。

2.5 小 结

细胞是生物体结构和功能的基本单位,对细胞分子调控网络的深入研究是揭开生命奥秘和征服疾病的关键途径。细胞在生理活动过程中会产生多种微量或痕量的活性小分子物质，如活性氧（ROS）、活性氮（RNS）、多巴胺、葡萄糖等，此类小分子物质表现出特殊的生理调控功能，对生命过程起着至关重要的影响。生物体内的活性氧在机体的生理与病理过程中具有信号传递调控、机体防御与应答等重要作用。它们在细胞的代谢过程中不断产生与消耗，参与控制细胞的增殖、分化与凋亡，并参与细胞的信号转导、多种因子生物学效应的启动。因此，对细胞中活性氧自由基的实时原位监测具有重要作用，它不仅有助于人们更好的理解和研究生物体内各个生理过程和病理过程的相关分子机制，还可以为许多疾病的诊断和治疗奠定基础。近年来，随着人们对生命科学研究的不断深入，对细胞内的重要小分子物质的研究越来越引人关注，原位、实时、定量检测细胞释放的活性小分子物质是研究其生物学功能的基础。因此，生物小分子传感器对于研究细胞活性小分子如何调节细胞的生物学功能、抗肿瘤药物的生物活性以及疾病的早期诊断等领域具有重要的理论指导意义。

参 考 文 献

[1] APEL K, HIRT H. Reactive oxygen species: Metabolism, oxidative stress, and signal transduction [J]. Annu Rev Plant Biol, 2004, 55: 373-399.

[2] SIES H, JONES D P. Reactive oxygen species (ROS) as pleiotropic physiological signalling agents [J]. Nat Rev Mol Cell Bio, 2020, 21 (7): 363-383.

[3] FORMAN H J, ZHANG H. Author Correction: Targeting oxidative stress in disease: Promise and limitations of antioxidant therapy [J]. Nat Rev Drug Discov, 2021, 20 (8): 652.

[4] HAYES J D, DINKOVA-KOSTOVA A T, TEW K D. Oxidative stress in cancer [J]. Cancer Cell, 2020, 38 (2): 167-197.

[5] SIES H, BELOUSOV V V, CHANDEL N S, et al. Defining roles of specific reactive oxygen species (ROS) in cell biology and physiology [J]. Nat Rev Mol Cell Biol, 2022, 23 (7): 499-515.

[6] COMMONER B, TOWNSEND J, PAKE G E. Free radicals in biological materials [J]. Nature, 1954, 174 (4432): 689-691.

[7] ALFADDA A A, SALLAM R M. Reactive oxygen species in health and disease [J]. J Biomed Biotechnol, 2012, 2012: 936486.

[8] BAYIR H. Reactive oxygen species [J]. Crit Care Med, 2005, 33 (12): S498-S501.

[9] OLMEZ I, OZYURT H. Reactive oxygen species and ischemic cerebrovascular disease [J]. Neurochem Int, 2012, 60 (2): 208-212.

[10] GALLEY H F. Oxidative stress and mitochondrial dysfunction in sepsis [J]. Brit J Anaesth, 2011, 107 (1): 57-64.

[11] URAKAWA H, KATSUKI A, SUMIDA Y, et al. Oxidative stress is associated with adiposity and insulin resistance in men [J]. J Clin Endocr Metab, 2003, 88 (10): 4673-4676.

[12] DATLA S R, GRIENDLING K K. Reactive oxygen species, NADPH oxidases, and hypertension [J]. Hypertension, 2010, 56 (3): 325-330.

[13] KIM G H, KIM J E, RHIE S J, et al. The role of oxidative stress in neurodegenerative diseases [J]. Exp Neurobiol, 2015, 24 (4): 325-340.

[14] SALMINEN A, OJALA J, KAARNIRANTA K, et al. Mitochondrial dysfunction and oxidative stress activate inflammasomes: Impact on the aging process and age-related diseases [J]. Cell Mol Life Sci, 2012, 69 (18): 2999-3013.

[15] LIOU G Y, STORZ P. Reactive oxygen species in cancer [J]. Free Radical Res, 2010, 44 (5): 479-496.

[16] FIRUZI O, MIRI R, TAVAKKOLI M, et al. Antioxidant therapy: Current status and future prospects [J]. Curr Med Chem, 2011, 18 (25): 3871-3888.

[17] VALKO M, LEIBFRITZ D, MONCOL J, et al. Free radicals and antioxidants in normal physiological functions and human disease [J]. Int J Biochem Cell Biol, 2007, 39 (1): 44-84.

[18] TURRENS J F. Mitochondrial formation of reactive oxygen species [J]. J Physiol-London, 2003, 552 (2): 335-344.

[19] ZONG W X, RABINOWITZ J D, WHITE E. Mitochondria and cancer [J]. Mol Cell, 2016, 61 (5): 667-676.

[20] LIANG H, LIU H, TIAN B, et al. Carbon quantum dot @ silver nanocomposite-based fluorescent imaging of intracellular superoxide anion [J]. Microchimica Acta, 2020, 187 (9): 484.

[21] STARKOV A A, FISKUM G, CHINOPOULOS C, et al. Mitochondrial α-ketoglutarate dehydrogenase complex generates reactive oxygen species [J]. J Neurosci, 2004, 24 (36): 7779-7788.

[22] JIE Z S, LIU J, SHU M C, et al. Detection strategies for superoxide anion: A review [J]. Talanta, 2022, 236: 122892.

[23] XU C, ZHANG W, WANG R X, et al. Versatile gold-coupled te-carbon dots for quantitative monitoring and efficient scavenging of superoxide anions [J]. Analytical Chemistry, 2021, 93 (26): 9111-9118.

[24] XIAO H B, ZHANG W, LI P, et al. Versatile fluorescent probes for imaging the superoxide

anion in living cells and in vivo [J]. Angew Chem Int Edit, 2020, 59 (11): 4216-4230.

[25] ZHANG J, YU Q H, CHEN W Y. Advancements in small molecule fluorescent probes for superoxide anion detection: A Review [J]. J Fluoresc, 2024.

[26] YIN J, LIU M F, REN W K, et al. Effects of dietary supplementation with glutamate and aspartate on diquat-induced oxidative stress in piglets [J]. Plos One, 2015, 10 (4): e0122893.

[27] EMERIT J, MICHELSON A M. Free-radicals in medicine and biology [J]. Semaine Des Hopitaux, 1982, 58 (45): 2670-2675.

[28] RUBBO H, RADI R, TRUJILLO M, et al. Nitric-oxide regulation of superoxide and peroxynitrite-dependent lipid-peroxidation-formation of novel nitrogen-containing oxidized lipid derivatives [J]. Journal of Biological Chemistry, 1994, 269 (42): 26066-26075.

[29] ZHANG Y, YUAN R, CHAI Y Q, et al. Simultaneous voltammetric determination for DA, AA and NO_2^- based on graphene/poly-cyclodextrin/MWCNTs nanocomposite platform [J]. Biosensors & Bioelectronics, 2011, 26 (9): 3977-3980.

[30] JENSEN F B, ROHDE S. Comparative analysis of nitrite uptake and hemoglobin-nitrite reactions in erythrocytes: Sorting out uptake mechanisms and oxygenation dependencies [J]. Am J Physiol-Reg I, 2010, 298 (4): R972-R982.

[31] DONG M, WANG C, DEEN W M, et al. Absence of 2′-deoxyoxanosine and presence of abasic sites in DNA exposed to nitric oxide at controlled physiological concentrations [J]. Chem Res Toxicol, 2003, 16 (9): 1044-1055.

[32] ALI A A, COULTER J A, OGLE C H, et al. The contribution of N_2O_3 to the cytotoxicity of the nitric oxide donor DETA/NO: An emerging role for S-nitrosylation [J]. Bioscience Rep, 2013, 33: 333-449.

[33] BREDT D S, WANG T L, COHEN N A, et al. Cloning and expression of 2 brain-specific inwardly rectifying potassium channels [J]. P Natl Acad Sci USA, 1995, 92 (15): 6753-6757.

[34] WANG Y, MARSDEN P A. Nitric oxide synthases: Gene structure and regulation [J]. Advances in pharmacology (San Diego, Calif), 1995, 34: 71-90.

[35] CAI H, HARRISON D G. Endothelial dysfunction in cardiovascular diseases—The role of oxidant stress [J]. Circ Res, 2000, 87 (10): 840-844.

[36] ZHANG Z, NAUGHTON D, WINYARD P G, et al. Generation of nitric oxide by a nitrite reductase activity of xanthine oxidase: A potential pathway for nitric oxide formation in the absence of nitric oxide synthase activity [J]. Biochem Bioph Res Co, 1998, 249 (3): 767-772.

[37] IGNARRO L J, BYRNS R E, BUGA G M, et al. Endothelium-derived relaxing factor from pulmonary-artery and vein possesses pharmacological and chemical-properties identical to those of nitric-oxide radical [J]. Circ Res, 1987, 61 (6): 866-879.

[38] CHAMPION H C, KADOWITZ P J. Vasodilator responses to acetylcholine, bradykinin, and substance P are mediated by a TEA-sensitive mechanism [J]. Am J Physiol-Reg I, 1997, 273 (1): R414-R422.

[39] WANG Y, HU S. A novel nitric oxide biosensor based on electropolymerization poly (toluidine blue) film electrode and its application to nitric oxide released in liver homogenate [J]. Biosensors & Bioelectronics, 2006, 22 (1): 10-17.

[40] VALKO M, LEIBFRITZ D, MONCOL J, et al. Free radicals and antioxidants in normal physiological functions and human disease [J]. Int J Biochem Cell B, 2007, 39 (1): 44-84.

[41] KOWALTOWSKI A J, DE SOUZA-PINTO N C, CASTILHO R F, et al. Mitochondria and reactive oxygen species [J]. Free Radical Biology and Medicine, 2009, 47 (4): 333-343.

[42] ARMSTRONG J S, WHITEMAN M. Measurement of reactive oxygen species in cells and mitochondria [J]. Method Cell Biol, 2007, 80: 355-377.

[43] GRANGE P A, CHEREAU C, RAINGEAUD J, et al. Production of superoxide anions by keratinocytes initiates P. acnes-induced inflammation of the skin [J]. PLoS Pathogens, 2009, 5 (7): e1000527.

[44] WENNMALM A, LANNE B, PETERSSON A S. Detection of endothelial-derived relaxing factor in human plasma in the basal state and following ischemia using electron-paramagnetic resonance spectrometry [J]. Analytical Biochemistry, 1990, 187 (2): 359-363.

[45] NILSSON R, PICK F M, BRAY R C, et al. Esr evidence for O2-as a long-lived transient in irradiated oxygenated alkaline aqueous solutions [J]. Acta Chem Scand, 1969, 23 (7): 2554-2556.

[46] KOHNO M. Applications of electron spin resonance spectrometry for reactive oxygen species and reactive nitrogen species research [J]. J Clin Biochem Nutr, 2010, 47 (1): 1-11.

[47] BORBAT P P, COSTA-FILHO A J, EARLE K A, et al. Electron spin resonance in studies of membranes and proteins [J]. Science, 2001, 291 (5502): 266-269.

[48] SKATCHKOV M P, SPERLING D, HINK U, et al. Quantification of superoxide radical formation in intact vascular tissue using a Cypridina luciferin analog as an alternative to lucigenin [J]. Biochem Bioph Res Co, 1998, 248 (2): 382-386.

[49] LU C, SONG G, LIN J M. Reactive oxygen species and their chemiluminescence-detection methods [J]. TrAC-Trends in Analytical Chemistry, 2006, 25 (10): 985-995.

[50] ROYALL J A, ISCHIROPOULOS H. Evaluation of 2′,7′-dichlorofluorescin and dihydrorhodamine

123 as fluorescent-probes for intracellular H_2O_2 in cultured endothelial-cells [J]. Arch Biochem Biophys, 1993, 302 (2): 348-355.

[51] OTTONELLO L, FRUMENTO G, ARDUINO N, et al. Immune complex stimulation of neutrophil apoptosis: Investigating the involvement of oxidative and nonoxidative pathways [J]. Free Radical Biology and Medicine, 2001, 30 (2): 161-169.

[52] GAO M Q, GUO S B, CHEN X H, et al. Molecular mechanisms of polypeptide from Chlamys farreri protecting HaCaT cells from apoptosis induced by UVA plus UVB [J]. Acta Pharmacol Sin, 2007, 28 (7): 1007-1014.

[53] LI J J, TANG Q, LI Y, et al. Role of oxidative stress in the apoptosis of hepatocellular carcinoma induced by combination of arsenic trioxide and ascorbic acid [J]. Acta Pharmacol Sin, 2006, 27 (8): 1078-1084.

[54] CHEN W L, QIAN Y, MENG W F, et al. A novel marine compound xyloketal B protects against oxidized LDL-induced cell injury in vitro [J]. Biochem Pharmacol, 2009, 78 (8): 941-950.

[55] LIU L L, SHENG B Y, YAN Y F, et al. Protective effect of anthocyanin against the oxidative stress in neuroblastoma N2a cells [J]. Prog Biochem Biophys, 2010, 37 (7): 779-785.

[56] GUO C X, ZHENG X T, LU Z S, et al. Biointerface by cell growth on layered graphene-artificial peroxidase-protein nanostructure for in situ quantitative molecular detection [J]. Adv Mater, 2010, 22 (45): 5164-5167.

[57] GUO C X, NG S R, KHOO S Y, et al. RGD-peptide functionalized graphene biomimetic live-cell sensor for real-time detection of nitric oxide molecules [J]. ACS Nano, 2012, 6 (8): 6944-6951.

[58] LUO Y P, TIAN Y, RUI Q. Electrochemical assay of superoxide based on biomimetic enzyme at highly conductive TiO_2 nanoneedles: From principle to applications in living cells [J]. Chemical Communications, 2009 (21): 3014-3016.

[59] MALINSKI T, TAHA Z. Nitric-oxide release from a single cell measured insitu by a porphyrinic-based microsensor [J]. Nature, 1992, 358 (6388): 676-678.

[60] MATSUMOTO F, TOKUDA K, OHSAKA T. Electrogeneration of superoxide ion at mercury electrodes with a hydrophobic adsorption film in aqueous media [J]. Electroanalysis, 1996, 8 (7): 648-653.

[61] CHEN J, WOLLENBERGER U, LISDAT F, et al. Superoxide sensor based on hemin modified electrode [J]. Sensors and Actuators B-Chemical, 2000, 70 (1/2/3): 115-120.

[62] GE B, LISDAT F. Superoxide sensor based on cytochrome c immobilized on mixed-thiol SAM with a new calibration method [J]. Anal Chim Acta, 2002, 454 (1): 53-64.

[63] DENG Z F, TIAN Y, YIN X, et al. Physical vapor deposited zinc oxide nanoparticles for direct electron transfer of superoxide dismutase [J]. Electrochem Commun, 2008, 10 (5): 818-820.

[64] WIGHTMAN R M. Probing cellular chemistry in biological systems with microelectrodes [J]. Science, 2006, 311 (5767): 1570-1574.

[65] GIL D M A, REBELO M J F. Evaluating the antioxidant capacity of wines: A laccase-based biosensor approach [J]. Eur Food Res Technol, 2010, 231 (2): 303-308.

[66] FERNANDEZ-PACHON M S, VILLANO D, GARCIA-PARRILLA M C, et al. Antioxidant activity of wines and relation with their polyphenolic composition [J]. Anal Chim Acta, 2004, 513 (1): 113-118.

[67] POPRAC P, JOMOVA K, SIMUNKOVA M, et al. Targeting free radicals in oxidative stress-related human diseases [J]. Trends Pharmacol Sci, 2017, 38 (7): 592-607.

[68] HAN M, LEE D, LEE S H, et al. Oxidative stress and antioxidant pathway in allergic rhinitis [J]. Antioxidants (Basel), 2021, 10 (8): 1266.

[69] LEE E R, KIM J H, CHOI H Y, et al. Cytoprotective effect of eriodictyol in UV-irradiated keratinocytes via phosphatase-dependent modulation of both the p38 MAPK and Akt signaling pathways [J]. Cell Physiol Biochem, 2011, 27 (5): 513-524.

[70] KUBO N, MORITA M, NAKASHIMA Y, et al. Oxidative DNA damage in human esophageal cancer: Clinicopathological analysis of 8-hydroxydeoxyguanosine and its repair enzyme [J]. Dis Esophagus, 2014, 27 (3): 285-293.

[71] NAKABEPPU Y, TSUCHIMOTO D, ICHINOE A, et al. Biological significance of the defense mechanisms against oxidative damage in nucleic acids caused by reactive oxygen species—From mitochondria to nuclei [J]. Ann Ny Acad Sci, 2004, 1011: 101-111.

[72] HOUSTIS N, ROSEN E D, LANDER E S. Reactive oxygen species have a causal role in multiple forms of insulin resistance [J]. Nature, 2006, 440 (7086): 944-948.

[73] CABISCOL E, TAMARIT J, ROS J. Oxidative stress in bacteria and protein damage by reactive oxygen species [J]. International microbiology: The official journal of the spanish society for microbiology, 2000, 3 (1): 3-8.

[74] HAYASHI T, FUKUTO J M, IGNARRO L J, et al. Gender differences in atherosclerosis-possible role of nitric-oxide [J]. Journal of Cardiovascular Pharmacology, 1995, 26 (5): 792-802.

[75] MALINSKI T, KAPTURCZAK M, DAYHARSH J, et al. Nitric-oxide synthase activity in genetic-hypertension [J]. Biochem Bioph Res Co, 1993, 194 (2): 654-658.

[76] TRACHOOTHAM D, ALEXANDRE J, HUANG P. Targeting cancer cells by ROS-mediated

mechanisms: A radical therapeutic approach? [J]. Nature Reviews Drug Discovery, 2009, 8 (7): 579-591.

[77] YANG J C, LU M C, LEE C L, et al. Selective targeting of breast cancer cells through ROS-mediated mechanisms potentiates the lethality of paclitaxel by a novel diterpene, gelomulide K [J]. Free Radical Biology and Medicine, 2011, 51 (3): 641-657.

[78] GUO C X, ZHENG X T, NG S R, et al. In situ molecular detection of ischemic cells by enhanced protein direct electron transfer on a unique horseradish peroxidase-Au nanoparticles-polyaniline nanowires biofilm [J]. Chemical Communications, 2011, 47 (9): 2652-2654.

[79] CHANDLER J M, COHEN G M, MACFARLANE M. Different subcellular distribution of Caspase-3 and Caspase-7 following fas-induced apoptosis in mouse liver [J]. Journal of Biological Chemistry, 1998, 273 (18): 10815-10818.

[80] SIMON H U, HAJ-YEHIA A, LEVI-SCHAFFER F. Role of reactive oxygen species (ROS) in apoptosis induction [J]. Apoptosis, 2000, 5 (5): 415-418.

[81] LEE Y S, SOHN K C, KIM K H, et al. Role of protein kinase C delta in X-ray-induced apoptosis of keratinocyte [J]. Experimental Dermatology, 2009, 18 (1): 50-56.

[82] STARKOV A A. The role of mitochondria in reactive oxygen species metabolism and signaling [J]. Ann Ny Acad Sci, 2008, 1147: 37-52.

[83] WANG X D. The expanding role of mitochondria in apoptosis [J]. Gene Dev, 2001, 15 (22): 2922-2933.

[84] GREEN D R, REED J C. Mitochondria and apoptosis [J]. Science, 1998, 281 (5381): 1309-1312.

[85] HOCKENBERY D, NUNEZ G, MILLIMAN C, et al. Bcl-2 is an inner mitochondrial-membrane protein that blocks programmed cell-death [J]. Nature, 1990, 348 (6299): 334-336.

[86] MCDONNELL T J, DEANE N, PLATT F M, et al. Bcl-2-immunoglobulin transgenic mice demonstrate extended B-cell survival and follicular lymphoproliferation [J]. Cell, 1989, 57 (1): 79-88.

[87] DU C Y, FANG M, LI Y C, et al. Smac, a mitochondrial protein that promotes cytochrome c-dependent caspase activation by eliminating IAP inhibition [J]. Cell, 2000, 102 (1): 33-42.

[88] THORNBERRY N A, LAZEBNIK Y. Caspases: Enemies within [J]. Science, 1998, 281 (5381): 1312-1316.

[89] HUANG P, FENG L, OLDHAM E A, et al. Superoxide dismutase as a target for the selective killing of cancer cells [J]. Nature, 2000, 407 (6802): 390-395.

[90] VALKO M, RHODES C J, MONCOL J, et al. Free radicals, metals and antioxidants in oxidative stress-induced cancer [J]. Chem-Biol Interact, 2006, 160 (1): 1-40.

[91] FRUEHAUF J P, MEYSKENS F L. Reactive oxygen species: A breath of life or death? [J]. Clin Cancer Res, 2007, 13 (3): 789-794.

[92] KALYANARAMAN B, DARLEY-USMAR V, DAVIES K J A, et al. Measuring reactive oxygen and nitrogen species with fluorescent probes: Challenges and limitations [J]. Free Radical Biology and Medicine, 2012, 52 (1): 1-6.

[93] BONFOCO E, KRAINC D, ANKARCRONA M, et al. Apoptosis and necrosis-2 distinct events induced, respectively, by mild and intense insults with N-methyl-D-aspartate or nitric-oxide superoxide in cortical cell-cultures [J]. P Natl Acad Sci USA, 1995, 92 (16): 7162-7166.

[94] LOBNER D. Comparison of the LDH and MTT assays for quantifying cell death: Validity for neuronal apoptosis? [J]. J Neurosci Meth, 2000, 96 (2): 147-152.

[95] MARINO M L, FAIS S, DJAVAHERI-MERGNY M, et al. Proton pump inhibition induces autophagy as a survival mechanism following oxidative stress in human melanoma cells [J]. Cell Death Dis, 2010, 1 (10): e87.

[96] WEINSTEIN I B, JOE A. Oncogene addiction versus oncogene amnesia: Perhaps more than just a bad habit? -Response [J]. Cancer Res, 2008, 68 (9): 3086.

[97] JONES R G, THOMPSON C B. Tumor suppressors and cell metabolism: A recipe for cancer growth [J]. Gene Dev, 2009, 23 (5): 537-548.

[98] HALLIWELL B. Oxidative stress and cancer: Have we moved forward? [J]. Biochem J, 2007, 401: 1-11.

[99] BARNESE K, GRALLA E B, CABELLI D E, et al. Manganous phosphate acts as a superoxide dismutase [J]. Journal of the American Chemical Society, 2008, 130 (14): 4604-4606.

[100] WANG M Q, YE C, BAO S J, et al. Controlled synthesis of Mn (PO) hollow spheres as biomimetic enzymes for selective detection of superoxide anions released by living cells [J]. Microchimica Acta, 2017, 184 (4): 1177-1184.

[101] DING A L, WANG B, MA X Q, et al. DNA-induced synthesis of biomimetic enzyme for sensitive detection of superoxide anions released from live cell [J]. Rsc Adv, 2018, 8 (22): 12354-12359.

[102] WANG Y, WANG D, SUN L H, et al. Constructing high effective nano-Mn(PO)-chitosan in situ electrochemical detection interface for superoxide anions released from living cell [J]. Biosensors & Bioelectronics, 2019, 133: 133-140.

[103] CUI M, REN J J, WEN X F, et al. Electrochemical detection of superoxide anion released by

living cells by manganese (Ⅲ) tetraphenyl porphine as superoxide dismutase mimic [J]. Chem Res Chinese U, 2020, 36 (5): 774-780.

[104] YUAN L, LIU S L, TU W W, et al. Biomimetic superoxide dismutase stabilized by photopolymerization for superoxide anions biosensing and cell monitoring [J]. Analytical Chemistry, 2014, 86 (10): 4783-4790.

[105] OLEAN-OLIVEIRA A, PACHECO J C, SERAPHIM P M, et al. Synergistic effect of reduced graphene oxide/azo-polymer layers on electrochemical performance and application as nonenzymatic chemiresistor sensors for detecting superoxide anion radicals [J]. Journal of Electroanalytical Chemistry, 2019, 852: 113520.

[106] CAI X, CHEN H L, WANG Z X, et al. 3D graphene-based foam induced by phytic acid: An effective enzyme-mimic catalyst for electrochemical detection of cell-released superoxide anion [J]. Biosensors & Bioelectronics, 2019, 123: 101-107.

[107] LIU L, ZHAO H L, SHI L B, et al. Enzyme-and metal-free electrochemical sensor for highly sensitive superoxide anion detection based on nitrogen doped hollow mesoporous carbon spheres [J]. Electrochim Acta, 2017, 227: 69-76.

[108] KLUN U, ZORKO D, STOJANOV L, et al. Amperometric sensor for gaseous H_2O_2 based on copper redox mediator incorporated electrolyte [J]. Sensors and Actuators Reports, 2023, 5: 100144.

[109] RAO D J, ZHANG J, ZHENG J B. Synthesis of silver nanoparticles-decorated $FePO_4$ nanosphere at a gas-liquid interface for the electrochemical detection of hydrogen peroxide [J]. J Chem Sci, 2016, 128 (5): 839-847.

[110] PENG L J, ZHOU H Y, ZHANG C Y, et al. Study on the peroxidase-like activity of cobalt phosphate and its application in colorimetric detection of hydrogen peroxide [J]. Colloid Surface A, 2022, 647: 129031.

[111] WANG M Q, YE C, BAO S J, et al. Nanostructured cobalt phosphates as excellent biomimetic enzymes to sensitively detect superoxide anions released from living cells [J]. Biosensors & Bioelectronics, 2017, 87: 998-1004.

[112] ZHAO X, PENG M H, LIU Y X, et al. Fabrication of cobalt nanocomposites as enzyme mimetic with excellent electrocatalytic activity for superoxide oxidation and cellular release detection [J]. ACS Sustain Chem Eng, 2019, 7 (12): 10227-10233.

[113] ZOU Z, CHEN J, SHI Z Z, et al. Cobalt phosphates loaded into iodine-spaced reduced graphene oxide nanolayers for electrochemical measurement of superoxide generated by cells [J]. Acs Appl Nano Mater, 2021, 4 (4): 3631-3638.

[114] ZOU Z, SHI Z A Z A, WU J G, et al. Atomically dispersed Co to an end-adsorbing molecule

for excellent biomimetically and prime sensitively detecting $O_2^{\cdot-}$ released from living cells [J]. Analytical Chemistry, 2021, 93 (31): 10789-10797.

[115] WANG Y, WANG M Q, LEI L L, et al. FePO$_4$ embedded in nanofibers consisting of amorphous carbon and reduced graphene oxide as an enzyme mimetic for monitoring superoxide anions released by living cells [J]. Microchimica Acta, 2018, 185 (2): 140.

[116] HU F X, GUO C X, YANG H B, et al. 3D Pt/Graphene foam bioplatform for highly sensitive and selective adsorption and detection of superoxide anions released from living cells [J]. Sensors and Actuators B-Chemical, 2019, 287: 209-217.

[117] YANG H X, HOU J G, WANG Z H, et al. An ultrasensitive biosensor for superoxide anion based on hollow porous PtAg nanospheres [J]. Biosensors & Bioelectronics, 2018, 117: 429-435.

[118] MADHURANTAKAM S, SELVARAJ S, RAYAPPAN J B B, et al. Exploring hesperidin-copper complex as an enzyme mimic for monitoring macrophage activity [J]. J Solid State Electr, 2018, 22 (6): 1893-1899.

[119] WU T D, LI L, SONG G J, et al. An ultrasensitive electrochemical sensor based on cotton carbon fiber composites for the determination of superoxide anion release from cells [J]. Microchimica Acta, 2019, 186 (3): 198.

[120] FAN W Z, LIU X H, WU J S, et al. Development of a novel silver-based sensing platform for detecting superoxide anion released from HeLa cells directly [J]. Electroanalysis, 2022, 34 (6): 987-994.

[121] ZHU D Z, HE P, KONG H, et al. Biomimetic graphene-supported ultrafine platinum nanowires for colorimetric and electrochemical detection of hydrogen peroxide [J]. Journal of Materials Chemistry B, 2022, 10 (44): 9216-9225.

[122] KO E, TRAN V K, SON S E, et al. Characterization of Au@PtNP/GO nanozyme and its application to electrochemical microfluidic devices for quantification of hydrogen peroxide [J]. Sensors and Actuators B-Chemical, 2019, 294: 166-176.

[123] ZHANG Y H, GUO C X, DU H, et al. Solvent-engineered morphologies of Mn-MOF toward ultrasensitive sensing cell superoxide [J]. Electrochim Acta, 2022, 431: 141147.

[124] LING P H, CHENG S, CHEN N, et al. Nanozyme-modified metal-organic frameworks with multienzymes activity as biomimetic catalysts and electrocatalytic interfaces [J]. ACS Appl Mater Inter, 2020, 12 (15): 17185-17192.

[125] WEI Z Q, LI W, YANG H, et al. Synthesis of 3D Co-based zeolitic imidazolate framework and application as electrochemical sensor for HO detection [J]. Int J Electrochem Sc, 2022, 17 (11): 221132.

[126] LIU X, XIANG M H, ZHANG X Y, et al. An enzyme-free electrochemical H_2O_2 sensor based on a nickel metal-organic framework nanosheet array [J]. Electroanalysis, 2022, 34 (2): 369-374.

[127] YANG X L, QIU W, GAO R W, et al. MIL-47 (V) catalytic conversion of H_2O_2 for sensitive H_2O_2 detection and tumor cell inhibition [J]. Sensors and Actuators B-Chemical, 2022, 354: 131201.

[128] LI J, XIE J, GAO L, et al. Au nanoparticles-3D graphene hydrogel nanocomposite to boost synergistically in situ detection sensitivity toward cell-released nitric oxide [J]. ACS Appl Mater Interfaces, 2015, 7 (4): 2726-2734.

[129] MAYER B, HEMMENS B. Biosynthesis and action of nitric oxide in mammalian cells [J]. Trends in Biochemical Sciences, 1997, 22 (12): 477-481.

[130] DEZFULIAN C, RAAT N, SHIVA S, et al. Role of the anion nitrite in ischemia-reperfusion cytoprotection and therapeutics [J]. Cardiovascular Research, 2007, 75 (2): 327-338.

[131] SHAH P, ZHU X, ZHANG X, et al. Microelectromechanical system-based sensing arrays for comparative in vitro nanotoxicity assessment at single cell and small cell-population using electrochemical impedance spectroscopy [J]. ACS Appl Mater Interfaces, 2016, 8 (9): 5804-5812.

[132] KIM M Y, NAVEEN M H, GURUDATT N G, et al. Detection of nitric oxide from living cells using polymeric zinc organic framework-derived zinc oxide composite with conducting polymer [J]. Small (Weinheim an der Bergstrasse, Germany), 2017, 13 (26): 1700502.

[133] XU H, LIAO C, LIU Y, et al. Iron phthalocyanine decorated nitrogen-doped graphene biosensing platform for real-time detection of nitric oxide released from living cells [J]. Anal Chem, 2018, 90 (7): 4438-4444.

[134] LIU Y L, JIN Z H, LIU Y H, et al. Stretchable electrochemical sensor for real-time monitoring of cells and tissues [J]. Angew Chem Int Ed Engl, 2016, 55 (14): 4537-4541.

[135] ZHAO X, WANG K, LI B, et al. Fabrication of a flexible and stretchable nanostructured gold electrode using a facile ultraviolet-irradiation approach for the detection of nitric oxide released from cells [J]. Anal Chem, 2018, 90 (12): 7158-7163.

3 电化学生物传感器在药物开发中的应用

3.1 引　　言

电化学是物理化学的一个分支,主要研究电与化学反应之间的关系,具有仪器简单、成本适中、便携等优点。早在 1800 年伏特就发明了第一个化学电池,到今天,它已经经历了 200 多年的发展。电化学领域的最新发展趋势涉及电化学分析、化学供电、电化学合成、光电化学等[1,2]。电化学已成为一种强大的分析技术,用于活细胞、生物活性分子和代谢物的化学分析。电化学生物传感器、微流体和质谱是电化学检测和监测中最常用的方法,它们构成了一系列非常有用的测量工具,可用于生物学和医学的各个领域。最近,电化学已被证明与纳米技术和基因工程相结合,以产生新的使能技术,提供快速、选择性和敏感的检测与诊断平台。在此将主要围绕应用电化学策略的效用及其与其他药物代谢和发现方法的结合,还讨论了电化学在药物研究中的当前挑战和可能的未来发展和应用。

无酶电化学生物传感器具有环境适用性强、稳定性高等特点[3,4]。纳米材料,尤其是金属纳米粒子,在改善传感器性能,提高非酶传感器灵敏度方面扮演着非常重要的角色,是材料科学中最基础、最活跃的组成部分。纳米技术的发展也为开发创新性的无酶电化学生物传感器带来了新的机遇和灵感。不同类型的纳米材料,如金属及其氧化物、碳纳米管、石墨烯及其衍生物、碳量子点、MOFs 基材料、MXenes 材料,以及纳米复合材料,已广泛应用于电化学生物传感器[5-8]。纳米材料具有类酶活性,表现出类似天然酶的酶促反应动力学和催化机理[9-11]。例如,类葡萄糖氧化酶活性、类过氧化物酶活性、类过氧化氢酶活性、类超氧化物歧化酶活性。与传统电化学生物传感器材料相比,纳米材料能够增强界面吸附性能,增加电催化活性,并促进电子转移动力。材料组合和制备方法可以有效地优化纳米材料性能,如表面电子迁移率和密度、协同效应、电荷载流子类型、表面电荷和带隙。基于纳米复合材料构建的合成仿生酶作为一种低成本的人工催化剂

可替代生物酶在 $O_2^{·-}$ 的检测中的作用功能。综上可知,构建非酶型电化学生物传感器用于 $O_2^{·-}$ 的检测将是未来的发展趋势。

电化学有许多不同的用途,其应用范围从生物医学化学到临床诊断。电子传递是指生物体氧化还原反应中的电子移动[12,13]。通常,电化学电池测量选择一个由三个电极组成的系统,包括工作电极、对电极和参比电极,这些电极浸泡在含有目标化合物和支持电解质的溶液中(见图 3.1(a))。这些电极连接到一个恒电位器,它控制工作电极的电位并测量产生的电流。工作电极,也称为研究电

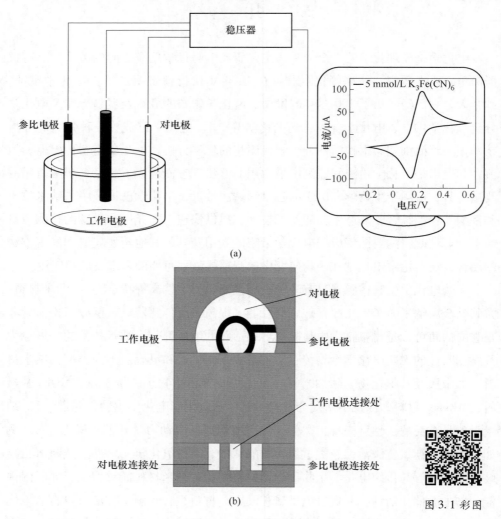

图 3.1 三电极检测系统

(a) 传统三电极系统的电化学检测平台示意图;(b) 丝网印刷电极示意图

极,与被分析物接触,并以受控的方式施加所需的电位,以促进电荷与被分析物之间的转移。参比电极相对于施加到工作电极的电位提供固定电位。计数(辅助)电极用于伏安分析或其他预计有电流流过的反应[14]。利用三电极系统进行基本分析,给出了材料的氧化/还原、电阻值和可逆性细节信息。作为传统电极的替代品,丝网印刷电极(SPEs)(见图 3.1(b))是通过在各种类型的塑料或陶瓷基板上印刷不同的油墨而生产的器件,由于其低成本和简单的制备工艺,使研究人员产生了极大的兴趣[15]。根据修饰工作电极的材料类型,SPE 可分为未修饰 SPE、膜修饰 SPE、酶修饰 SPE 和抗原/抗体修饰 SPE[16]。SPEs 的多功能性在于电极可以被广泛地修改。与三电极系统不同,双电极恒电位器只允许使用工作电极和反电极,而更通用的四电极恒电位器使用两个工作电极[17]。

3.2 生物学研究中的电化学参数与技术

生命现象的许多过程都伴随着电子转移反应,因此电化学方法是研究生物系统中电子转移及相关过程较好地揭示生命本质的方法。在医学研究中,电化学为检测药物代谢物对生物分子的反应性提供了一种经济、快速、清洁的系统。例如,甲基乙二醛是人血浆中的生物标志物,有许多方法报道了其测定方法。为了定量分析甲基乙二醛,研制了一种单壁碳纳米管修饰的玻碳电极[18]。在 0.1 ~ 100 μmol/L 的宽线性范围内,该电化学传感器对人血浆甲基乙二醛具有明显的还原峰,灵敏度高达 76.3 nA/(μmol·L^{-1})。因此,这种有效的系统有助于实验室检测甲基乙二醛并阐明其在糖尿病相关并发症中的作用。Johnson 等人证明了方波电位脉冲可以显著提高 Pt 电极上各种有机化合物的电化学检测[19]。Nouri-Nigjeh 与他的同事展示了电化学氧化药物代谢的好处,通过方波电位脉冲氧化利多卡因[20]。利用方波电位脉冲,通过优化循环次数和电位,可以提高药物代谢产物的电化学生产选择性和产率。因此,电化学方法经常用于获取药物分子及其作用机制的关键信息,以发现有价值的新药先导化合物[21,22]。

特别是伏安法和电解法更适用于研究电活性药物、药物代谢物以及代谢物与生物分子[12]的相互作用。循环伏安法(CV)和线性扫描伏安法(LSV)[21]技术在过去几年中得到了极大的普及,因为它们可以可靠地评估电极过程和氧化还原机制[23]。溶出伏安法(SV)由于能够对预浓缩分析物进行超灵敏检测而得到广泛应用[24-26]。安培法已应用于电生理学研究囊泡释放事件[27,28]。差分脉冲伏安法（DPV）和方波伏安法（SWV）等其他技术在测定药物和生物样品中痕量电活性化合物方面尤为重要[29-31]。

3.3　电化学方法在体外评价药物代谢中的应用

药物代谢的研究对药物的成功开发至关重要[32]。药物化合物可在体内转化为治疗活性或毒性代谢物。虽然药物代谢和处置的研究需要实验模型，但正确选择和应用正确的检测技术对候选药物的决策和成功推进至关重要。近20年来，能够快速评价和模拟药物代谢氧化反应的电化学技术得到了发展，并广泛应用于电活性药物反应机理的研究[33]，包括电化学生物传感器，电化学与微流体和质谱（MS）相结合。电化学生物传感器可用于监测药物代谢物及其生物活化途径，电化学质谱（EC-MS）常被用于药物元素分析和配置，电化学微流体在药物疗效和药动学性质评价方面具有优势。由 Nouri-Nigjeh 及其同事提出，亚秒方波电位脉冲通过电化学氧化促进非那西丁转化为对乙酰氨基酚[34]。该方法是在药物发现和开发的早期阶段模拟药物氧化代谢的一种有潜力的分析技术。近年来，电化学阻抗谱（EIS）以其灵敏度高、无标记等特点在介电性能测量中受到越来越多的关注[35,36]。

3.4　电化学生物传感器在药物代谢中的研究

生物或生化过程的量化是生物学、生物医学和生物技术应用的最大挑战。电化学生物传感器的功能是将生物反应转化为电信号，通过检测、传输和记录有关生理或生化变化的信息，为分析生物样品或药物的含量提供了一种有吸引力的手段[37]。与其他电化学传感器相比，电化学生物传感器具有特异性好、灵敏度高、易于制作等优点[22,38]。典型的生物传感器包括样品、生物受体、传感器和信号传输与处理系统。生物识别元件通常包含酶、抗体、细胞、组织或微生物。换能器包括电化学、光学和压电元件。信号传输和信号处理系统是数据采集和处理的两个重要组成部分。该生物传感器的识别元件可以确定对靶标药物的选择性。而传感器的灵敏度受传感器本身的影响较大通常作为电化学传感器性能的评价标准[39,40]。

近年来,基于纳米材料分析方法的发展已引起众多应用的关注,包括基础生物学研究、健康监测、临床诊断、药物分析、食品安全和环境监测等。纳米材料由于其出色的理化特性——较大的比表面积,较强的吸附能力和反应能力,已成为潜在的分析探针,不仅提供了更高的灵敏度,而且还为分析单分子领域提供了新的平台。此外,某些纳米颗粒表现出强大的仿生电催化活性为开发具有更高稳定性的新型超灵敏和高选择性无酶生物传感器提供了巨大的可能性。由纳米材料修饰电极制备的电化学生物传感器平台因其易于操作、经济高效、高灵敏度和选择性、快速响应、精巧便携等优点而被广泛用作强大的分析方法。金属纳米粒子和碳纳米材料等具有良好导电性能,可以加快生物分子与电极之间的电子传递速率。纳米材料还具有良好的生物相容性,可以将大量的生物分子固定在电极表面,并使其保持原有的生物活性。纳米材料的修饰使得电化学传感器的性能得到了巨大的改善。

3.4.1 酶标记的电化学传感器在药物代谢物检测中的应用

近 20 年来,生物传感器的相关研究经历了爆炸式的发展,目前电化学生物传感器已广泛应用于模拟药物代谢和代谢物对生物分子的反应。在生物传感器结合实验中,已有几种方法成功地将生物识别分子固定在传感表面上,并具有完整的功能,例如基于酶的电化学生物传感器,它通过记录电子传递信号来研究代谢反应,是一种有效的工具[41]。细胞色素 P450(CYP)酶是一个单加氧酶超家族,在葡萄糖氧化酶和细胞色素 C 之后具有重要地位[42,43]。它们是肝脏中主要的药物代谢酶,因为它们可以代谢目前市场上 75%~90% 的药物。Sheila 等人报道,同时使用第二种药物抑制 CYP 介导的药物代谢导致人类使用生物电化学的药物相互作用[44]。这一观察结果表明,CYP3A4 作为 CYP 的同工酶,在药物代谢方面发挥着最重要的作用。Victoria 和他的同事通过基于生物电催化的潜在底物或 CYP 抑制剂筛选,分析了 CYP 的电流-电压特性、电催化循环的化学计量学、氧化还原热力学和过氧化物分流反应[45]。CYP 电化学与纳米技术相结合,是一种新型的高通量筛选方法,可以缩短分析时间并减少分析步骤,降低研究成本。Alexey 等人利用高灵敏度电化学方法评价了噁唑啉基衍生物的活性[46]。这些噁唑啉基衍生物作为潜在的 CYP17A1 抑制剂,通过与 CYP17A1 结合抑制前列腺癌细胞的生长和增殖,对新药的开发具有重要意义[46]。为了避免无效治疗带来的不必要的费用,人们一直致力于开发治疗药物监测(TDM)的新技术。然而,由于传统的分析技术（如色谱和免疫测定），TDM 仅限于少数设施。Baj-Rossi 等人利用多壁碳纳米管（MWCNTs）作为在生物传感中增强电子转移的非常有前途的纳米材料，介绍了一种使用 MWCNTs 和微粒体 CYP1A2 作为电极材料的石墨 SPEs[47]，这为开发全面运作的治疗性生物传感器提供了一种新方法。因此，由于电化学生物传感器的便携性、高灵敏度和极端简单性，TDM 实践可以极大地受益于电化学生物传感器。由于 CYP 催化的反应是一种常见的实例，CYP 酶膜被用来修饰电极，可以作为一种筛选工具来研究药物代谢的电催化活性[48-50]。为了了解各种碳电极材料与人肝微粒体（HLM）界面的直接电化学性质，Walgama 等人开发了一种新型微粒体生物反应器，将 HLM 直接吸附在抛光的基面热解石墨（BPG）、边缘面热解石墨（EPG）、玻碳（GC）或高纯度石墨（HPG）电极上。考虑到 HLM 在药物开发和毒理学领域的重要性，这种生物分析平台将对廉价的药物代谢和抑制分析非常有用[51]。迄今为止，更多的努力集中在酶催化循环的直接电子传递上，这有助于生物传感器的开发，而不需要氧化还原转移蛋白和辅助因子。

3.4.2 纳米酶标记的电化学传感器在药物研究中的应用

在电化学传感器的构建过程中，电极修饰材料在整个传感器中起到关键作用，因此，选择一种合适的电极修饰材料是十分重要的。典型的电化学生物传感器含有生物识别元件（酶、蛋白质、抗体、核酸、组织或受体），从而选择性地与靶向分析物发生电化学反应，并产生可测量电信号。然而，利用酶的特异性检测细胞代谢物时，往往受限于酶易失活、导电性差或环境要求苛刻等因素，在传感器的分析检测时会产生大的背景电流、弱的电信号或慢的电子传递速率、酶的生物传感器成本高，长期稳定性差，因为天然酶在不利的环境中容易变性而失去生物活性，严重限制其快速发展。基于上述原因，仿生酶作为一种既具有酶催化活性又具有良好导电性能、稳定物理化学性能和生物相容性的新型电极材料成为传感领域的研究热点[52]。大多数仿生酶将碳纳米材料作为基底材料，使用电化学还原、原位沉积、高温煅烧等制备方法，合成高性能的纳米复合电极材料仿生酶。纳米复合电极材料仿生酶电化学传感器通过进行氧化还原反应实现对于 RONS 等细胞生物小分子的电化学检测，这对于预防疾病与药物发现具有重要意义。

超氧阴离子（$O_2^{·-}$）作为 ROS 家族中最活泼的成员之一，其广泛存在于各种生物系统，并参与多种重要的生物学过程[53]。研究已表明 $O_2^{·-}$ 与多种生物现象和氧化应激相关疾病有关，例如衰老、癌症发展进程、动脉粥样硬化和神经退行性疾病的发病机制等[54-56]。在正常生理条件下，$O_2^{·-}$对生物体有益，并在信号转导过程中起到重要作用。然而，当身体受到某些环境因素或其他外部条件的影响时，$O_2^{·-}$的产生和去除可能会偏离平衡，从而导致过量的$O_2^{·-}$产生。然而，过量积累的$O_2^{·-}$可能会对生物膜、组织和生物体产生一定的损伤，并诱发衰老和各种疾病，如心脑血管疾病、糖尿病、肿瘤、神经退行性疾病、动脉硬化和癌症等[57]。因此，快速、可靠、灵敏、特异地检测$O_2^{·-}$对于药物研究、疾病诊断和健康筛查具有重要意义。由于$O_2^{·-}$可以通过多种非催化或/和酶促反应途径迅速歧化，致使其具有寿命短、活性高等特点，导致$O_2^{·-}$的检测极其困难[58]。目前，常用于检测$O_2^{·-}$的方法主要有电子自旋共振法、分光光度法、化学发光法和电化学技术[59-62]，其中电化学技术因其具有操作简单、灵敏度高、实时检测、仪器成本低等优点，已受到研究者广泛的关注。迄今为止，各种各样的电化学酶传感器已经被报道，利用特定的酶催化剂，如细胞色素C和超氧化物歧化酶（SOD）[63,64]。基于纳米材料的氧化还原酶型$O_2^{·-}$电化学传感器进一步提高了氧化还原酶的活性与传感器的性能[65]。电化学检测$O_2^{·-}$的方法非常有吸引力，但依赖于超氧化物歧化酶，其存在成本高、耐用性差的问题。纳米科学的进步使得构建一个具有高灵敏度和特异性的纳米级仿生传感平台成为可能。锰可以用作有效的催化剂，早在1982年的相关报道表明其具有保护体内超氧化物的作用[66]。后续研究发现仅有磷酸锰（$Mn_3(PO_4)_2$）能够催化$O_2^{·-}$的歧化作用，而Mn^{2+}自由基离子只与$O_2^{·-}$发生化学反应[67]。据报道，一种电化学传感器通过将Mn^{2+}结合到高导电性的二氧化钛纳米针上，以纳米聚合物作为黏合剂来检测$O_2^{·-}$ [68]。在我们前期研究工作中，通过磷酸锰自组装合成了一种仿生酶$Mn_3(PO_4)_2$纳米片并进一步组装在碳纳米管（CNTs）上形成独特的CNT/DNA@$Mn_3(PO_4)_2$纳米仿生酶复合材料（见图3.2），其中$Mn_3(PO_4)_2$能有效催化$O_2^{·-}$的畸变，而碳纳米管能快速催化$O_2^{·-}$的畸变发生电子转移，从而实现对$O_2^{·-}$的高灵敏度和特异性检测，具有长期稳定性[69]。该仿生酶$O_2^{·-}$传感器进一步用于监测在药物刺激作用下小鼠癌细胞和正常皮肤细胞原位释放$O_2^{·-}$水平，表现出优异的实时定量检测能力。这项工作展示一种性能与天然酶相当的仿生酶为基础的生物传感器，同时具有更高的耐用性，因此在基础领域有着广阔的应用前景。

超氧阴离子（$O_2^{·-}$）与一氧化氮（NO）在许多生理与病理性疾病中发挥着重

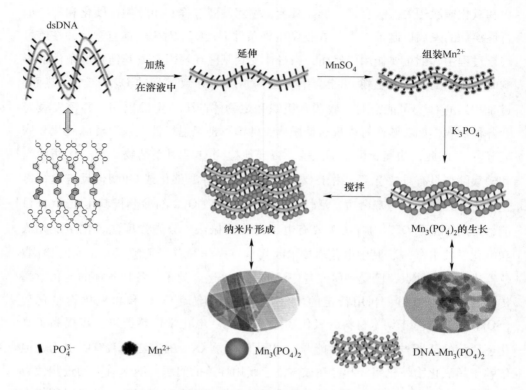

图 3.2 CNT/DNA@$Mn_3(PO_4)_2$ 纳米片的合成示意图

要的作用[70,71]。NO 的过度存在可以引起亚硝基化反应，这一反应能够改变蛋白质的结构和正常的生理功能[8]。超氧阴离子与一氧化氮可以相互作用产生大量的过氧亚硝酸盐，它是一个非常重要的氧化剂具有强大的氧化能力[72-75]。由于超氧阴离子与一氧化氮具有重要的生物学意义，因此研究者一直致力于用不同的方法研究它们的重要作用，例如荧光探针标记法[76-78]、化学发光法[79]和电子自旋共振法[80]。然而，这些方法具有冗长乏味的操作程序和昂贵的仪器使用费用。目前，电化学生物传感器由于制作容易、高灵敏度和特异性好已经引起极大的关注。Guo 制备了石墨烯/过氧化氢人工酶/蛋白质的纳米复合结构的生物相容界面电极原位定量检测细胞释放的活性氧。在我们前期的工作中，合成了 CNT/DNA@$Mn_3(PO_4)_2$ 纳米复合材料电化学传感器对肿瘤细胞释放的超氧阴离子的实时、灵敏和特异性的检测[81]。这些研究表明电化学生物传感器用于实时检测溶液中活性氧浓度不受细胞内新陈代谢和调控途径的干扰，因此该方法是一种非常有希望的检测方法。

由于 RONS 参与了许多信号通路包括 MAP 激酶/ERK 通路，研究发现 PLX4032 诱导的细胞生长抑制过程中发挥重要作用[82,83]。Meenhard Herlyn 研究表明 PLX4032 在治疗时可能导致肿瘤代谢从糖酵解到氧化磷酸化的转移[84]，研究表明 PLX4032 可能具有损伤细胞氧化磷酸化的能力，目前研究已表明细胞的氧化磷酸的损伤会导致细胞中活性氧水平上升，引起细胞氧化损伤。研究通过电化学生物传感器与生物学相结合的方法，揭示了 PLX4032 诱导黑色素瘤细胞发生氧化损伤导致细胞凋亡的作用，超氧阴离子与一氧化氮电化学生物传感器详细分析了 PLX4032 处理人黑色素瘤细胞后超氧阴离子与一氧化氮的释放水平[85]。索拉非尼是一种新型多靶点抗肿瘤药物，不仅通过阻断由 RAF/MEK/ERK 介导的细胞信号传导通路而直接抑制肿瘤细胞的增殖，还可通过抑制 VEGFR 和血小板衍生生长因子（PDGF）受体而阻断肿瘤新生血管的形成，间接地抑制肿瘤细胞的生长[86]。索拉非尼是晚期肝细胞癌患者全身治疗的优先选择药物。但是，随着治疗时间的延长索拉非尼出现的耐药问题一定程度上阻碍了该药物在临床治疗中的应用[87]。通过 CNT/DNA@ $Mn_3(PO_4)_2$ 纳米仿生酶修饰的电化学传感器探讨了索拉非尼 FGF19 在肝癌索拉非尼耐药过程中的 $O_2^{·-}$ 和 NO 水平进行作用，辅助检测发现，肝癌细胞中 FGF19 的表达或者超活化 FGF19/FGFR4 信号通路是索拉非尼肝癌细胞耐药的主要作用机制之一[88,89]。CYT997（Lexibulin）是一种微管抑制剂，它通过有效的抑制微管聚合和阻断细胞微管发挥其潜在的细胞毒性和血管阻断能力[90]。通过 CNT/DNA@ $Mn_3(PO_4)_2$ 纳米仿生酶复合材料修饰的电化学生物传感器分别检测了对照组、HCQ 处理组、CYT997 处理组、HCQ 与 CYT997 联合处理组中 HNSCC 细胞（HN12）产生 ROS 和 $O_2^{·-}$，由实验可知，它们联合使用与单独使用相比可以引起更高水平的 $O_2^{·-}$，上述数据表明该传感器在监测 CYT997 触发的 HNSCC 细胞释放的 $O_2^{·-}$ 水平中发挥重要作用，进而为揭示 CYT997 诱导的 HNSCC 细胞发生凋亡提供参考依据。

综上可知，电化学生物传感器通过监测细胞代谢物进而对抗肿瘤药物作用效果进行评价，将为开发新的小分子药物提供新的途径。

3.5 电化学微流控技术在药物代谢中的应用

微流体技术有可能彻底改变传统的细胞生物学方法[91]，并能更好地检测微环境下细胞的生长情况。微流体技术在工程、生物、医学等领域有着广泛的应用前景[92]。早期的微流控概念可以追溯到 19 世纪 70 年代在硅片上利用光刻技术产生的气相色谱仪，后来发展为微流控毛细管电泳和微反应器。微流体的一个重要特征是具有独特性质的微尺度流体环境，如层流和液滴。利用这些独特的流体现象，可以用常规方法实现难以完成一系列微加工和微操作的微流控。目前，微流控技术与电化学技术相结合，通过与生物重要分子的复杂电化学相互作用，提高了实验数据质量的自动测定，减少了分析时间和成本[93,94]。

电微流体是一个相对年轻的领域，但在生物医学研究中被认为具有很大的发展潜力和广阔的前景[95-98]。它在药物代谢方面取得了实质性的成功。例如，连续流电合成已被用于检测几种商业药物的代谢物的各种化学反应性，如芳香羟基化、烷基氧化和谷胱甘肽偶联[99]。在微流控结构中使用电化学，每小时测量 10 ~ 100 mg 的纯分离代谢物。Gu 等人建立了一种简单而稳定的浓度梯度微流控系统，用于测定乙酰胆碱酯酶的反应酶抑制浓度[100]。与传统的酶抑制分析方法相比，该系统的样品消耗减少了约 1000 倍，适用于许多生化反应（如药物筛选和动力学研究），只要其中一种反应物或产物具有电化学活性。多巴胺是自主神经系统中一种重要的神经递质，与奖赏、情绪和成瘾以及帕金森病和亨廷顿病等神经系统疾病有关[101]。目前，最流行的体内多巴胺释放监测方法不可能同时测量多巴胺及其代谢物，这使得研究药物或其他治疗对多巴胺代谢途径的影响成为一项挑战。采用微芯片电泳的分离速度快（亚分钟），效率高，并且需要低样本量（从 pL 到 nL）[102]，它已发展为分离和检测分析物在体内多巴胺代谢途径[103]。基于电泳的微芯片由 5 μm 聚二甲基硅氧烷分离通道组成，结构简单，可为特定代谢途径化合物的检测和脑微透析样品的在线监测打开新的窗口。

3.6 化学-质谱联用检测药物代谢物

药物开发早期阶段的常规代谢研究包括以肝细胞和细胞提取物为基础的体外试验以及动物模型试验。模拟药物氧化代谢的可能性使 EC-MS 联合使用的吸引力大大增加[104,105]。电化学因此受益于结构信息的识别和反应产物的定量在纳摩尔水平通过质谱的手段[106]。质谱与电化学的结合始于 20 世纪 70 年代初。该领域的早期研究旨在检测特定药物在特定条件下的反应产物。最近的出版物报道了不同溶液条件（pH 值和电解质）和电化学辅助芬顿反应的影响的比较[107,108]。

Oberacher 等人使用基于在线 EC-MS 的方法测定了 $(C_5Me_5)_2Mo_2O_5$ 和相关

配合物的还原，并成功地鉴定了单核到四核的有机金属钼氧化物。这项工作可能为新的催化氧化还原化学和金属配合物分析的发展开辟道路[109]。Jahn等人利用电喷雾电离（ESI)-MS和电感耦合等离子体质谱（ICPMS）对候选药物进行了定性和定量分析[110]。在本研究中，为了鉴定代谢物，补充测量是不可避免的，并揭示了n-脱烷基是胺碘酮和托瑞米芬的主要代谢反应。EC-MS也可用于阐明分子结构。例如，已鉴定出两种姜黄素类似物（命名为NC2067和NC2081），其结构骨架为1,5-二芳基-3-氧-1,4-戊二烯，并使用EC-MS对其诊断产物离子进行了定性和定量分析[111]。

3.7 电化学方法在抗肿瘤药物发现中的应用

从药物发现到商业化是一个昂贵、漫长和渐进的过程，这一过程促进了药物发现早期阶段越来越有选择性、可靠和快速的高通量筛选（HTS）检测的发展。基于细胞的HTS检测提供了候选药物的毒性特征和疗效的早期指示。如果我们把活细胞看作一个电化学系统，氧化还原反应引起的电子产生和电荷转移以及活细胞中离子组成和浓度的变化可以用来表征细胞在均相溶液中的活力[112,113]。基于细胞的生物传感器集成了基于细胞活性和功能、细胞屏障行为或记录/刺激电致细胞电位的生物识别元件和电化学转导单元[112]。与其他方法相比，这些生物传感器具有响应分析物提供生理学相关数据和检测分析物生物利用度的能力，具有更高的生物催化活性、更低的生产成本和更强的稳定性[114,115]。

抗癌药物在消除癌细胞、缩小肿瘤、防止癌症转移扩散、缓解各种癌症症状等方面具有一定的功能。随着科学技术的飞速发展，各种抗癌药物被开发出来，这些药物的持续生产和常规临床应用需要即用型定性和定量分析平台。紫杉醇是从太平洋紫杉树皮中的天然化合物中提取出来的，它被认为是治疗卵巢癌、乳腺癌和非小细胞肺癌的有效药物。研究报道了一种紫杉醇阴极溶出SWV分析方法，采用电活性表面积为$0.0278/cm^2$的汞滴电极在碱条件下检测紫杉醇[116]。研究还发现基于PGEs的DPV分析方法可用于开发和改进紫杉醇传感器，其中紫杉醇通过其与鲑鱼精子双链DNA（ds-DNA）的相互作用来确定。在$0.5 \sim 1.3 \text{ V}$的电位范围内，相互作用降低了鸟嘌呤和腺嘌呤氧化信号的强度，从而在$0.2 \sim 10.0 \text{ μmol/L}$的线性范围内实现紫杉醇电分析[117]。Tajik等人的另一项研究用碳/金属氧化物/聚合物（ds-DNA-MWCNT-TiO_2/ZrO_2-壳聚糖）纳米复合材料修饰了

PGE 表面。紫杉醇具有良好的电子转移性质和较高的活性位点，从而提高了紫杉醇的伏安氧化反应。该合成的纳米复合电极为紫杉醇的检测提供了一个灵敏、稳定、选择性的检测平台，线性范围为 0.7 nmol/L ~ 1.87 μmol/L，检出限为 0.01 nmol/L[118]。阿霉素是一种从母链霉菌中提取的蒽环类药物，用于治疗各种类型的白血病、肺癌和乳腺癌。研究通过碳盘工作电极快速灵敏地分析阿霉素的方法，电化学催化背后的机制是醋酸缓冲液中的伪可逆反应，随着 pH 值的增加，一个明确的单峰向负电位移动证实了这一点[119]。临床和环境应用是这些传感器设计的最终目的。为此，Nicole 及其同事开发了一种具有与靶分子结合能力的人工单链核酸（Aptassensor）检测阿霉素的技术，可以在 4 ~ 1000 nmol/L 的宽浓度范围内进行分析检测，其中适配体在金电极上的饱和浓度在 125 nmol/L[120]。此外，开发检测两种不同的药物的传感器将有助于节省传感器的设计成本和检测时间。例如，试图开发出同时检测阿霉素和另一种抗癌药物（甲氨蝶呤）的药物在生物、临床和制药领域的分析与应用。在 GCE 上合成了一种环糊精-石墨烯杂化纳米复合材料（CD-GN/GCE），用于同时检测阿霉素和甲氨蝶呤。与在普通 GCE 上获得的结果相比，CD-GN/GCE 上的阿霉素和甲氨蝶呤的峰值电流分别增加了 26.5 倍和 23.7 倍[121]。另一种基于 DNA/贵金属（DNA/银）的传感器，制造用于使用固态伏安法检测阿霉素，通过银 NPs 与阿霉素偶联前后的 TEM 图像证实了银 NPs 在偶联后的尺寸保持不变。在标记过程中使用（红线）~17 和（黑线）~1 的阿霉素每 AgNP 载药量在 0.3 mol/L KCl 下测得的伏安结果的比较表明，由于阿霉素的插层不利，灵敏度随着阿霉素载药量的增加而降低[122]。伊马替尼（IMA）是一种治疗慢性髓性白血病（CML）和胃肠道间质瘤（GIST）的抗癌药物。需要在血液、脑脊液、血浆和尿液等生物液体中监测该药。基于上述要求，我们开发了一种树突层状中空纤维氧化石墨烯传感器，用于生物样品中 IMA 的同时微提取和电化学检测[123]。采用尼罗替尼、波纳替尼和达沙替尼三种抗癌药物对 IMA 具有高选择性，检出限为 7.39 nmol/L，进行干扰研究。SPCE 被用于制造传感器，因为众所周知，它们可以实现简单、快速激活和低成本的传感器，用于许多药物的定量检测[124]。例如，使用 SPCE（CNT）实现了 IMA 的伏安检测，并在 0.75 V 的峰值电位下显示了 IMA 在哌嗪环上的氧化行为。这种电极机制表明，催化过程是不可逆的，从 0.71 V 的电流显著增加可以看出，没有进一步的还原，并遵循典型的吸附控制过程。此外，将该设计的电极与铂和金工作电极进行了比较，两种电极都无法达到 SPCE 检测 IMA 的功效。在治疗免疫功能低下的

3.7 电化学方法在抗肿瘤药物发现中的应用

患者时，可口服、肌肉或静脉给药。在这种情况下，一种药物可能会抑制另一种药物的活性，这可能导致严重的副作用或可能需要改变剂量。为了避免这些并发症，同时进行药物检测是可取的。采用金属氧化物/碳基电极（NiO-ZnO/MWCNT-COOH/GCE）同时检测抗癌药物（IMA）和抗真菌药物（伊曲康唑）。DPV 结果表明，伊曲康唑的氧化峰电流增加，LOD 为 4.1 nmol/L，灵敏度为 2.64 μA/(μmol·L^{-1}·cm^2)。IMA 的线性范围为 0.015~2.0 μmol/L，LOD 为 2.4 nmol/L，灵敏度为 9.64 μA/(μmol·L^{-1}·cm^2)[125]。最近，Tseng 等人提出了一种基于氧化石墨烯片氧化钛钙（GOS/CaTiO$_3$）的纳米催化剂，用于电化学检测氟他胺，氟他胺是一种用于治疗前列腺癌和白血病的非甾体抗雄激素。图 3.3(a)~(c) 为通过 XRD、TEM 和 EDX 表面表征得到的纳米催化剂的形貌、结构和组成。图 3.3(d) 和 (e) 显示了对氟他胺的 i-t 响应和相关校准图，显示出宽线性范围（0.015~1184 μmol/L），低检测限（5.7 nmol/L）和高灵敏度（1.073 μA/(μmol·L^{-1}·cm^2)）。该传感器进一步与人类血液和尿液样本进行了测试，证实了该传感器在抗癌药物检测中的应用前景[126]。

在药物发现中，专门的电化学策略已被用于确定药物的安全性和有效性。例如，Ding 等人开发了一种用于细胞固定化和电化学研究的仿生凝胶[127]。利用该

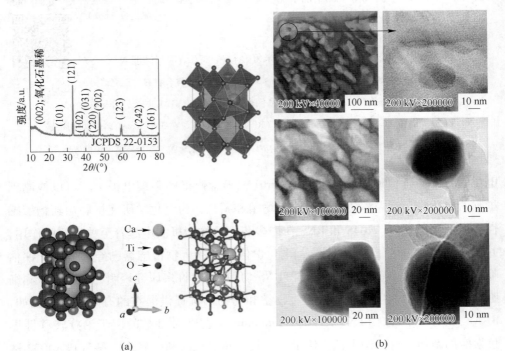

(a)　　　　　　　　　　　　(b)

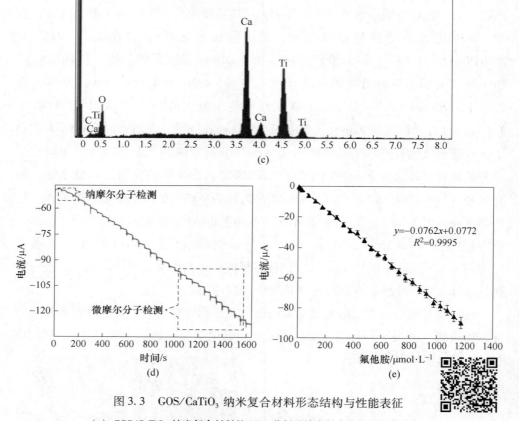

图 3.3　GOS/CaTiO₃ 纳米复合材料形态结构与性能表征

(a) GOS/CaTiO₃ 纳米复合材料的 XRD 分析及纳米复合材料的正交结构；
(b) GOS/CaTiO₃ 纳米材料的 TEM 图像；(c) GOS/CaTiO₃ 纳米复合材料的 EDS 分析；(d) GOS/CaTiO₃ 修饰 GCE（RPM = 1200）下，在 0.05 mol/L 磷酸盐缓冲液（pH = 7.0）中低浓度和高浓度氟他胺药物的 i-t 反应；(e) 氟他胺与阴极峰值电流的线性图[126]

图 3.3 彩图

电化学方法，将 K562 白血病细胞附着在纳米金修饰的碳糊电极上，可以检测到细胞的氧化峰。通过测定细胞黏附、增殖和凋亡，为电化学研究抗肿瘤药物敏感性提供了新的途径。ROS 和 RNS 在许多药物的作用机制中起着至关重要的作用，并被证明参与了电子转移[128-130]，电化学传感器和生物传感器已经发展用于评估临床和生理分析中感兴趣的 ROS 和 RNS 物种[131]。许多研究表明，电化学监测细胞自由基水平的效用可能为细胞生物学和药物筛选提供潜在的有效方法。例如，BRAF V600E 抑制剂 PLX4032（Vemurafenib）是 FDA 批准的用于治疗转移性黑色素瘤的新药，可抑制黑色素瘤细胞生长。超氧化物和一氧化氮参与 PLX4032 诱导的生长抑制已经通过电化学生物传感器确定[85]。在这项研究中，使用电化学传感器和常规荧光素染色技术监测 PLX4032 挑战的 BRAF V600E 突变体 A375 细胞释放的超氧自由基 $O_2^{·-}$ 和一氧化氮（NO）。有趣的是，科学家们还发现，PLX4032 处理降低了 BRAF V600E 突变体黑色素瘤细胞的线粒体膜电位。这表明电化学方法可以用于检测生物分子（包括脂质、蛋白质、DNA 和小化合物）诱导的早期细胞凋亡。

近年来，利用电微流控检测平台建立了基于单细胞的药物代谢物分析，可准确监测细胞对药物治疗等外部信号反应过程中的细胞间异质性[132,133]。Yeon 团队的研究展示了集成在微滴板底部的电化学阻抗传感器阵列的使用，以评估使用潜在细胞毒性药物治疗的细胞生长情况[134]。微加工细胞芯片系统为细胞毒性检测提供了一种简便、实时的监测工具。电池-衬底阻抗传感（ECIS）是另一项重要的电化学技术。ECIS 方法工作原理示意图如图 3.4 所示，作为细胞黏附的生物活性平台，传感器可以通过监测电流和电压变化来评估动态信息。当细胞失去附着力时，离子形式的电流可以自由地从表面流向电极。而电池附着在电极表面会阻碍自由电子流向电极表面，从而增加系统的电阻。因此，ECIS 可用于实时监测细胞黏附、扩散、移动等动态信息[135,136]。ECIS 已广泛应用于药物开发，如早期安全性评估和细胞代谢机制研究[137-140]。Michaelis 等人已将该技术应用于监测

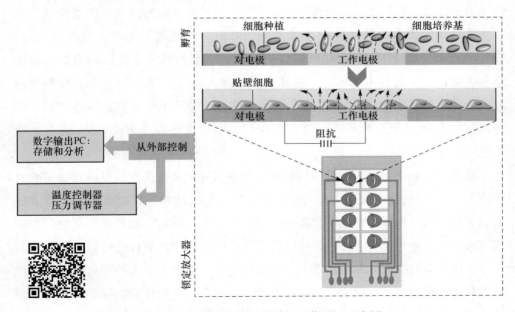

图 3.4 彩图　　图 3.4　电池-衬底阻抗传感器工作原理示意图

可溶性 RGDS 肽对建立的细胞层的干扰[141]。他们还研究了上皮 MDCK 细胞在不同蛋白涂层上的附着和扩散，并研究了二价阳离子对扩散动力学的影响[102]。Bennet 等人报道了一种新的基于 ECIS 的分析工具，用于实时测量视网膜神经节细胞的发光效应时间反应功能（RGC-5）[101]。ECIS 不仅测定了所分析药物（如β-胡萝卜素、槲皮素、胍丁氨酸、谷胱甘肽）对 RGC-5 细胞的保护作用，而且测定了无毒安全剂量下的最大药物活性，与标准生物学测量结果非常吻合。这些研究表明，电化学策略将在治疗学领域引起广泛的兴趣。

3.8 电化学生物传感器在药物开发中的局限性与挑战

电化学方法具有广阔的应用前景，但不能取代生物系统中传统的代谢研究。作为一种特别有用的代谢物标准物合成补充技术，缺乏现场测量，测量时间长，特异性和准确性较低仍然是一个挑战。例如，过氧化氢和抗坏血酸是困扰电化学检测 $O_2^{\cdot-}$ 的特别严重的干扰。因为它们是共存的生物化合物，广泛存在于生物系统中，并具有电化学氧化活性[69]。为了避免虚假信号和排除干扰，需要合适的电极修饰材料来提高电化学评价的特异性。另外，对于稳定的和反应性的代谢物，虽然电化学氧化可以模拟代谢途径，但相关方法并不适合作为预测生物系统中所有药物分子代谢命运的通用工具[142,143]。为了更好地探索和拓展电化学技术作为传统代谢研究在药物发现和开发中的补充，应在以下几个方面继续探索：（1）在线耦合电化学与固定化共轭酶；（2）用天然酶或不太复杂的模拟酶修饰电极表面；（3）评估电化学方法在还原性代谢方面的应用；（4）电化学氧化与互补的非生物系统相结合；（5）在电化学质谱小型化电池领域取得进展；（6）提高制备电化学微流控系统的商业可用性，能够提供高于 100 mg/h 的电化学产率[22]。

综上可知，电化学是传统体内或体外代谢研究的补充技术，可以在很短的时间内提供分子/药物的氧化代谢指纹图谱。与传统方法相比，电化学具有许多显著的优势，包括快速、敏感和清洁地生成和直接鉴定稳定的物种和代谢物。在临床试验前，各种电化学方法已被广泛应用于模拟药物代谢和代谢物对生物分子的反应。电化学与微流体和质谱等其他技术相结合，创造了强大的平台来模拟各种氧化和还原过程，从而促进了新药的开发。电化学、生物化学和医学知识可以整合为氧化还原选择性治疗和机制研究的发展制定策略。

3.9 小　　结

医学的持续目标是更好地治疗病人，这需要包括电化学在内的各个领域的大量研究。电化学已成为药物评估和发现的有力替代方法。目前，许多合适的电子生物分析工具和策略已经开发和验证。电化学生物传感器、EC-MS、电化学微流控等常用技术可用于分析药物的疗效和性质。此外，电化学方法可以应用于高通量药物筛选，所获得的分析信息的高可靠性使得加速药物发现的过程成为可能。虽然电化学方法不会完全取代传统的生物系统研究，但它有助于我们进一步研究生物反应途径和生物大分子的相互作用。电化学在药物研究中的全部潜力才刚刚开始被发现，在不久的将来，这一领域的重大新应用当然可以预期。

参 考 文 献

[1] GUL T, BISCHOFF R, PERMENTIER H P. Electrosynthesis methods and approaches for the preparative production of metabolites from parent drugs [J]. TrAC-Trends in Analytical Chemistry, 2015, 70: 58-66.

[2] SPEISER B. Numerical simulations in electrochemistry [J]. In: Encyclopedia of Applied Electrochemistry. edn.: Springer, 2014: 1380-1385.

[3] DERINA K, KOROTKOVA E, BAREK J. Non-enzymatic electrochemical approaches to cholesterol determination [J]. Journal of Pharmaceutical and Biomedical Analysis, 2020, 191: 113538.

[4] PANAHI Z, CUSTER L, HALPERN J M. Recent advances in non-enzymatic electrochemical detection of hydrophobic metabolites in biofluids [J]. Sensors and Actuators Reports, 2021, 3: 100051.

[5] THATIKAYALA D, PONNAMMA D, SADASIVUNI K K, et al. Progress of advanced nanomaterials in the non-enzymatic electrochemical sensing of glucose and H_2O_2 [J]. Biosensors (Basel), 2020, 10 (11): 151.

[6] GONÇALVES J M, DE FARIA L V, NASCIMENTO A B, et al. Sensing performances of spinel ferrites MFe_2O_4 (M = Mg, Ni, Co, Mn, Cu and Zn) based electrochemical sensors: A review [J]. Analytica chimica acta, 2022, 1233: 340362.

[7] ZHU H, LIANG X H, CHEN J T, et al. The influence of ionic liquids on the fabrication of nonenzymatic glucose electrochemical sensor [J]. Talanta, 2011, 85 (3): 1592-1597.

[8] LI B, SONG H Y, DENG Z P, et al. Novel sensitive amperometric hydrogen peroxide sensor using layered hierarchical porous α-MoO and GO modified glass carbon electrode [J]. Sensors and Actuators B-Chemical, 2019, 288: 641-648.

[9] CORMODE D P, GAO L, KOO H. Emerging biomedical applications of enzyme-like catalytic nanomaterials [J]. Trends Biotechnol, 2018, 36 (1): 15-29.

[10] WU J, WANG X, WANG Q, et al. Nanomaterials with enzyme-like characteristics (nanozymes): Next-generation artificial enzymes (Ⅱ) [J]. Chem Soc Rev, 2019, 48 (4): 1004-1076.

[11] SONG W, ZHAO B, WANG C, et al. Functional nanomaterials with unique enzyme-like characteristics for sensing applications [J]. J Mater Chem B, 2019, 7 (6): 850-875.

[12] BANDI V, GOBEZE H B, D'SOUZA F. Ultrafast photoinduced electron transfer and charge stabilization in donor-acceptor dyads capable of harvesting near-infrared light [J]. Chem-Eur J, 2015, 21 (32): 11483-11494.

[13] ZHOU F M, MILLHAUSER G L. The rich electrochemistry and redox reactions of the copper sites in the cellular prion protein [J]. Coordin Chem Rev, 2012, 256 (19/20): 2285-2296.

[14] SIMPSON P C, PETISCE J R, CARR-BRENDEL V E, et al. Electrode systems for electrochemical sensors [J]. Electroanalysis, 2013, 25 (1): 29-46.

[15] METTERS J P, KADARA R O, BANKS C E. New directions in screen printed electroanalytical sensors: An overview of recent developments [J]. Analyst, 2011, 136 (6): 1067-1076.

[16] RENEDO O D, ALONSO-LOMILLO M A, MARTINEZ M J A. Recent developments in the field of screen-printed electrodes and their related applications [J]. Talanta, 2007, 73 (2): 202-219.

[17] DRYDEN M D M, WHEELER A R. DStat: A versatile, open-source potentiostat for electroanalysis and integration [J]. Plos One, 2015, 10 (10): e0140349.

[18] CHATTERJEE S, WEN J L, CHEN A C. Electrochemical determination of methylglyoxal as a biomarker in human plasma [J]. Biosensors & Bioelectronics, 2013, 42: 349-354.

[19] HOEKSTRA J C, JOHNSON D C. Waveform optimization for integrated square-wave detection of biogenic amines following their liquid chromatographic separation [J]. Anal Chim Acta, 1999, 390 (1/2/3): 45-54.

[20] NOURI-NIGJEH E, PERMENTIER H P, BISCHOFF R, et al. Electrochemical oxidation by square-wave potential pulses in the imitation of oxidative drug metabolism [J]. Analytical Chemistry, 2011, 83 (14): 5519-5525.

[21] ÁLVAREZ-LUEJE A, PÉREZ M, ZAPATA C. Electrochemical methods for the in vitro

assessment of drug metabolism [J]. Topics on Drug Metabolism, 2012, 9: 221-247.

[22] JURVA U, WEIDOLF L. Electrochemical generation of drug metabolites with applications in drug discovery and development [J]. TrAC-Trends in Analytical Chemistry, 2015, 70: 92-99.

[23] MENG Y, ALDOUS L, BELDING S R, et al. The formal potentials and electrode kinetics of the proton/hydrogen couple in various room temperature ionic liquids [J]. Chemical Communications, 2012, 48 (45): 5572-5574.

[24] BATCHELOR-MCAULEY C, KÄTELHÖN E, BARNES E O, et al. Recent advances in voltammetry [J]. ChemistryOpen, 2015, 4 (3): 224-260.

[25] MENG Y, NORMAN S, HARDACRE C, et al. The electroreduction of benzoic acid: Voltammetric observation of adsorbed hydrogen at a platinum microelectrode in room temperature ionic liquids [J]. Physical Chemistry Chemical Physics, 2013, 15 (6): 2031-2036.

[26] CHEN A, ROGERS E I, COMPTON R G. Abrasive stripping voltammetry in room temperature ionic liquids [J]. Electroanalysis, 2009, 21 (1): 29-35.

[27] HAYAT A, CATANANTE G, MARTY J L. Current trends in nanomaterial-based amperometric biosensors [J]. Sensors, 2014, 14 (12): 23439-23461.

[28] YANG P, LI X, WANG L, et al. Sandwich-type amperometric immunosensor for cancer biomarker based on signal amplification strategy of multiple enzyme-linked antibodies as probes modified with carbon nanotubes and concanavalin A [J]. Journal of Electroanalytical Chemistry, 2014, 732: 38-45.

[29] SHAH B, CHEN A C. Novel electrochemical approach for the monitoring of biodegradation of phenolic pollutants and determination of enzyme activity [J]. Electrochem Commun, 2012, 25: 79-82.

[30] KHATER D Z, EL-KHATIB K M, HAZAA M M, et al. Development of bioelectrochemical system for monitoring the biodegradation performance of activated sludge [J]. Appl Biochem Biotech, 2015, 175 (7): 3519-3530.

[31] CHATTERJEE S, CHEN A C. Voltammetric detection of the alpha-dicarbonyl compound: Methylglyoxal as a flavoring agent in wine and beer [J]. Anal Chim Acta, 2012, 751: 66-70.

[32] KARLSSON F H, BOUCHENE S, HILGENDORF C, et al. Utility of in vitro systems and preclinical data for the prediction of human intestinal first-pass metabolism during drug discovery and preclinical development [J]. Drug Metabolism and Disposition, 2013, 41 (12): 2033-2046.

[33] FABER H, MELLES D, BRAUCKMANN C, et al. Simulation of the oxidative metabolism of diclofenac by electrochemistry/ (liquid chromatography) /mass spectrometry [J]. Anal

Bioanal Chem, 2012, 403 (2): 345-354.

[34] NOURI-NIGJEH E, BISCHOFF R, BRUINS A P, et al. Electrochemical oxidation by square-wave potential pulses in the imitation of phenacetin to acetaminophen biotransformation [J]. Analyst, 2011, 136 (23): 5064-5067.

[35] RANDVIIR E P, BANKS C E. Electrochemical impedance spectroscopy: An overview of bioanalytical applications [J]. Anal Methods-Uk, 2013, 5 (5): 1098-1115.

[36] HU Y F, ZUO P, YE B C. Label-free electrochemical impedance spectroscopy biosensor for direct detection of cancer cells based on the interaction between carbohydrate and lectin [J]. Biosensors & Bioelectronics, 2013, 43: 79-83.

[37] WANG J, WU C, HU N, et al. Microfabricated electrochemical cell-based biosensors for analysis of living cells in vitro [J]. Biosensors, 2012, 2 (2): 127-170.

[38] TURNER A P. Biosensors: Sense and sensibility [J]. Chemical Society Reviews, 2013, 42 (8): 3184-3196.

[39] D'ORAZIO P. Biosensors in clinical chemistry-2011 update [J]. Clinica Chimica Acta, 2011, 412 (19): 1749-1761.

[40] SUNDRAMOORTHY A K, GUNASEKARAN S. Applications of graphene in quality assurance and safety of food [J]. TrAC-Trends in Analytical Chemistry, 2014, 60: 36-53.

[41] YUAN T, PERMENTIER H, BISCHOFF R. Surface-modified electrodes in the mimicry of oxidative drug metabolism [J]. TrAC-Trends in Analytical Chemistry, 2011, 70: 50-57.

[42] RAVINDRANATH V, STROBEL H W. Cytochrome P450-mediated metabolism in brain: Functional roles and their implications [J]. Expert Opin Drug Met, 2013, 9 (5): 551-558.

[43] BOHMDORFER M, MAIER-SALAMON A, RIHA J, et al. Interplay of drug metabolizing enzymes with cellular transporters [J]. Wiener medizinische Wochenschrift (1946), 2014, 164 (21/22): 461-471.

[44] SADEGHI S J, FERRERO S, DI NARDO G, et al. Drug-drug interactions and cooperative effects detected in electrochemically driven human cytochrome P450 3A4 [J]. Bioelectrochemistry, 2012, 86: 87-91.

[45] SHUMYANTSEVA V V, BULKO T V, SUPRUN E V, et al. Electrochemical investigations of cytochrome P450 [J]. Biochimica Et Biophysica Acta-Proteins and Proteomics, 2011, 1814 (1): 94-101.

[46] KUZIKOV A V, DUGIN N O, STULOV S V, et al. Novel oxazolinyl derivatives of pregna-5, 17 (20)-diene as 17 alpha-hydroxylase/17, 20-lyase (CYP17A1) inhibitors [J]. Steroids, 2014, 88: 66-71.

[47] BAJ-ROSSI C, JOST T R, CAVALLINI A, et al. Continuous monitoring of Naproxen by a

cytochrome P450-based electrochemical sensor [J]. Biosensors & Bioelectronics, 2014, 53: 283-287.

[48] SCHNEIDER E, CLARK D S. Cytochrome P450 (CYP) enzymes and the development of CYP biosensors [J]. Biosensors and Bioelectronics, 2013, 39 (1): 1-13.

[49] WU Y H, LIU X Q, ZHANG L, et al. An amperometric biosensor based on rat cytochrome P450 1A1 for benzo [a] pyrene determination [J]. Biosensors & Bioelectronics, 2011, 26 (5): 2177-2182.

[50] SHUMYANTSEVA V V, MAKHOVA A A, BULKO T V, et al. Electrocatalytic cycle of P450 cytochromes: The protective and stimulating roles of antioxidants [J]. Rsc Adv, 2015, 5 (87): 71306-71313.

[51] WALGAMA C, NERIMETLA R, MATERER N F, et al. A simple construction of electrochemical liver microsomal bioreactor for rapid drug metabolism and inhibition assays [J]. Analytical chemistry, 2015, 87 (9): 4712-4718.

[52] ZOU Z, SHI Z Z, WU J G, et al. Atomically dispersed Co to an end-adsorbing molecule for excellent biomimetically and prime sensitively detecting $O_2^{\cdot-}$ released from living cells [J]. Analytical Chemistry, 2021, 93 (31): 10789-10797.

[53] ANDERSEN J K. Oxidative stress in neurodegeneration: Cause or consequence? [J]. Nat Med, 2004, 10 (Suppl): S18-S25.

[54] ZOU Z Z, CHANG H C, LI H L, et al. Induction of reactive oxygen species: An emerging approach for cancer therapy [J]. Apoptosis, 2017, 22 (11): 1321-1335.

[55] HAYASHI T, KATO N, FURUDOI K, et al. Early-life atomic-bomb irradiation accelerates immunological aging and elevates immune-related intracellular reactive oxygen species [J]. Aging Cell, 2023, 22 (10): e13940.

[56] KLAUNIG J E, KAMENDULIS L M. The role of oxidative stress in carcinogenesis [J]. Annu Rev Pharmacol Toxicol, 2004, 44: 239-267.

[57] CHANG K H, CHEN C M. The role of oxidative stress in parkinson's disease [J]. Antioxidants (Basel), 2020, 9 (7): 597.

[58] HWANG C, YOO J, JUNG G Y, et al. Biomimetic superoxide disproportionation catalyst for anti-aging lithium-oxygen batteries [J]. ACS Nano, 2019, 13 (8): 9190-9197.

[59] JIE Z, LIU J, SHU M, et al. Detection strategies for superoxide anion: A review [J]. Talanta, 2022, 236: 122892.

[60] LEE D K, JANG H D. Carnosic acid attenuates an early increase in ROS levels during adipocyte differentiation by suppressing translation of NOX4 and inducing translation of antioxidant enzymes [J]. Int J Mol Sci, 2021, 22 (11): 6096.

[61] LIU F X, JIANG X C, HE N, et al. Electrochemical investigation for enhancing cellular antioxidant defense system based on a superoxide anion sensor [J]. Sensors and Actuators B-Chemical, 2022, 368: 132190.

[62] KIMURA S, INOGUCHI T, YAMASAKI T, et al. A novel DPP-4 inhibitor teneligliptin scavenges hydroxyl radicals: In vitro study evaluated by electron spin resonance spectroscopy and in vivo study using DPP-4 deficient rats [J]. Metabolism, 2016, 65 (3): 138-145.

[63] LVOVICH V, SCHEELINE A. Amperometric sensors for simultaneous superoxide and hydrogen peroxide detection [J]. Anal Chem, 1997, 69 (3): 454-462.

[64] BATRA B, SANGWAN S, AHLAWAT J, et al. Electrochemical sensing of cytochrome using graphene oxide nanoparticles as platform [J]. Int J Biol Macromol, 2020, 165: 1455-1462.

[65] TIAN Y, MAO L, OKAJIMA T, et al. Superoxide dismutase-based third-generation biosensor for superoxide anion [J]. Anal Chem, 2002, 74 (10): 2428-2434.

[66] ARCHIBALD F S, FRIDOVICH I. The scavenging of superoxide radical by manganous complexes: In vitro [J]. Archives of biochemistry and biophysics, 1982, 214 (2): 452-463.

[67] BARNESE K, GRALLA E B, CABELLI D E, et al. Manganous phosphate acts as a superoxide dismutase [J]. Journal of the American Chemical Society, 2008, 130 (14): 4604-4606.

[68] LUO Y, TIAN Y, RUI Q. Electrochemical assay of superoxide based on biomimetic enzyme at highly conductive TiO_2 nanoneedles: From principle to applications in living cells [J]. Chemical communications (Cambridge, England), 2009 (21): 3014-3016.

[69] MA X Q, HU W H, GUO C X, et al. DNA-templated biomimetic enzyme sheets on carbon nanotubes to sensitively in situ detect superoxide anions released from cells [J]. Advanced Functional Materials, 2014, 24 (37): 5897-5903.

[70] GHASEMI M, FATEMI A. Pathologic role of glial nitric oxide in adult and pediatric neuroinflammatory diseases [J]. Neurosci Biobehav R, 2014, 45: 168-182.

[71] MURAD F. Discovery of some of the biological effects of nitric oxide and its role in cell signaling [J]. Bioscience Rep, 2004, 24 (4/5): 452-474.

[72] STARKOV A A. The role of mitochondria in reactive oxygen species metabolism and signaling [J]. Ann Ny Acad Sci, 2008, 1147: 37-52.

[73] LIAW N, FOX J M D, SIDDIQUI A H, et al. Endothelial nitric oxide synthase and superoxide mediate hemodynamic initiation of intracranial aneurysms [J]. Plos One, 2014, 9 (7): e101721.

[74] CARR A C, MCCALL M R, FREI B. Oxidation of LDL by myeloperoxidase and reactive nitrogen species-reaction pathways and antioxidant protection [J]. Arterioscl Throm Vas, 2000, 20 (7): 1716-1723.

[75] CALCERRADA P, PELUFFO G, RADI R. Nitric oxide-derived oxidants with a focus on peroxynitrite: Molecular targets, cellular responses and therapeutic implications [J]. Curr Pharm Design, 2011, 17 (35): 3905-3932.

[76] YU L, FAVOINO E, WANG Y Y, et al. The CSPG4-specific monoclonal antibody enhances and prolongs the effects of the BRAF inhibitor in melanoma cells [J]. Immunol Res, 2011, 50 (2/3): 294-302.

[77] MYHRE O, ANDERSEN J M, AARNES H, et al. Evaluation of the probes 2',7'-dichlorofluorescin diacetate, luminol, and lucigenin as indicators of reactive species formation [J]. Biochem Pharmacol, 2003, 65 (10): 1575-1582.

[78] PESHAVARIYA H M, DUSTING G J, SELEMIDIS S. Analysis of dihydroethidium fluorescence for the detection of intracellular and extracellular superoxide produced by NADPH oxidase [J]. Free Radical Res, 2007, 41 (6): 699-712.

[79] JIANG H, PARTHASARATHY D, TORREGROSSA A C, et al. Analytical techniques for assaying nitric oxide bioactivity [J]. J Vis Exp, 2012 (64): e3722.

[80] SERRANDER L, CARTIER L, BEDARD K, et al. NOX4 activity is determined by mRNA levels and reveals a unique pattern of ROS generation [J]. Biochem J, 2007, 406: 105-114.

[81] MA X, HU W, GUO C, et al. DNA-templated biomimetic enzyme sheets on carbon nanotubes to sensitively in situ detect superoxide anions released from cells [J]. Advanced Functional Materials, 2014, 24 (37): 5897-5903.

[82] IRANI K, GOLDSCHMIDT-CLERMONT P J. Ras, superoxide and signal transduction [J]. Biochem Pharmacol, 1998, 55 (9): 1339-1346.

[83] WU L Y, DE CHAMPLAIN J. Effects of superoxide on signaling pathways in smooth muscle cells from rats [J]. Hypertension, 1999, 34 (6): 1247-1253.

[84] ROESCH A, VULTUR A, BOGESKI I, et al. Overcoming intrinsic multidrug resistance in melanoma by blocking the mitochondrial respiratory chain of slow-cycling JARID1B (high) cells [J]. Cancer Cell, 2013, 23 (6): 811-825.

[85] YU L, GAO L X, MA X Q, et al. Involvement of superoxide and nitric oxide in BRAF V600E inhibitor PLX4032-induced growth inhibition of melanoma cells [J]. Integrative Biology, 2014, 6 (12): 1211-1217.

[86] GORES G J. Decade in review-hepatocellular carcinoma: HCC-subtypes, stratification and sorafenib [J]. Nat Rev Gastroenterol Hepatol, 2014, 11 (11): 645-647.

[87] FAN G, WEI X, XU X. Is the era of sorafenib over? A review of the literature [J]. Therapeutic Advances in Medical Oncology, 2020, 26 (12): 1758835920927602.

[88] GAO L, WANG X, TANG Y, et al. FGF19/FGFR4 signaling contributes to the resistance of

hepatocellular carcinoma to sorafenib [J]. J Exp Clin Cancer Res, 2017, 36 (1): 8.

[89] GAO L X, SHAY C, LV F L, et al. Implications of FGF19 on sorafenib-mediated nitric oxide production in hepatocellular carcinoma cells-A short report [J]. Cell Oncol, 2018, 41 (1): 85-91.

[90] BURNS C J, FANTINO E, PHILLIPS I D, et al. CYT997: A novel orally active tubulin polymerization inhibitor with potent cytotoxic and vascular disrupting activity in vitro and in vivo [J]. Mol Cancer Ther, 2009, 8 (11): 3036-3045.

[91] YIN H B, MARSHALL D. Microfluidics for single cell analysis [J]. Curr Opin Biotech, 2012, 23 (1): 110-119.

[92] SACKMANN E K, FULTON A L, BEEBE D J. The present and future role of microfluidics in biomedical research [J]. Nature, 2014, 507 (7491): 181-189.

[93] NGE P N, ROGERS C I, WOOLLEY A T. Advances in microfluidic materials, functions, integration, and applications [J]. Chemical Reviews, 2013, 113 (4): 2550-2583.

[94] GENCOGLU A, MINERICK A R. Electrochemical detection techniques in micro-and nanofluidic devices [J]. Microfluidics and Nanofluidics, 2014, 17 (5): 781-807.

[95] NEUŽI P, GISELBRECHT S, LÄNGE K, et al. Revisiting lab-on-a-chip technology for drug discovery [J]. Nature Reviews Drug discovery, 2012, 11 (8): 620-632.

[96] GHAEMMAGHAMI A M, HANCOCK M J, HARRINGTON H, et al. Biomimetic tissues on a chip for drug discovery [J]. Drug Discovery Today, 2012, 17 (3): 173-181.

[97] KEPP O, GALLUZZI L, LIPINSKI M, et al. Cell death assays for drug discovery [J]. Nature Reviews Drug Discovery, 2011, 10 (3): 221-237.

[98] BRESLIN S, O'DRISCOLL L. Three-dimensional cell culture: The missing link in drug discovery [J]. Drug Discovery Today, 2013, 18 (5): 240-249.

[99] STALDER R, ROTH G P. Preparative microfluidic electrosynthesis of drug metabolites [J]. Acs Medicinal Chemistry Letters, 2013, 4 (11): 1119-1123.

[100] CORAZAO-ROZAS P, GUERRESCHI P, JENDOUBI M, et al. Mitochondrial oxidative stress is the achille's heel of melanoma cells resistant to Braf-mutant inhibitor [J]. Oncotarget, 2013, 4 (11): 1986-1998.

[101] BENNET D, KIM S. Impedance-based cell culture platform to assess light-induced stress changes with antagonist drugs using retinal cells [J]. Analytical Chemistry, 2013, 85 (10): 4902-4911.

[102] MICHAELIS S, WEGENER J, ROBELEK R. Label-free monitoring of cell-based assays: Combining impedance analysis with SPR for multiparametric cell profiling [J]. Biosensors and Bioelectronics, 2013, 49: 63-70.

[103] SAYLOR R A, REID E A, LUNTE S M. Microchip electrophoresis with electrochemical detection for the determination of analytes in the dopamine metabolic pathway [J]. Electrophoresis, 2015, 36 (16): 1912-1919.

[104] VAN DEN BRINK F T, OLTHUIS W, VAN DEN BERG A, et al. Miniaturization of electrochemical cells for mass spectrometry [J]. TrAC-Trends in Analytical Chemistry, 2015, 87 (3): 1527-1535.

[105] VAN DEN BRINK F T, BÜTER L, ODIJK M, et al. Mass spectrometric detection of short-lived drug metabolites generated in an electrochemical microfluidic chip [J]. Analytical Chemistry, 2015, 87 (3): 1527-1535.

[106] CHEN H, ZHANG Y H, MUTLIB A E, et al. Application of on-line electrochemical derivatization coupled with high-performance liquid chromatography electrospray ionization mass spectrometry for detection and quantitation of (p-chlorophenyl) aniline in biological samples [J]. Analytical Chemistry, 2006, 78 (7): 2413-2421.

[107] RODRIGO M, OTURAN N, OTURAN M. Electrochemically assisted remediation of pesticides in soils and water: A review [J]. Chemical Reviews, 2014, 114 (17): 8720-8745.

[108] YAVUZ Y, SHAHBAZI R, KOPARAL A S, et al. Treatment of basic red 29 dye solution using iron-aluminum electrode pairs by electrocoagulation and electro-fenton methods [J]. Environmental Science and Pollution Research, 2014, 21 (14): 8603-8609.

[109] OBERACHER H, PITTERL F, ERB R, et al. Mass spectrometric methods for monitoring redox processes in electrochemical cells [J]. Mass Spectrometry Reviews, 2015, 34 (1): 64-92.

[110] JAHN S, LOHMANN W, BOMKE S, et al. A ferrocene-based reagent for the conjugation and quantification of reactive metabolites [J]. Anal Bioanal Chem, 2012, 402 (1): 461-471.

[111] AWAD H, DAS U, DIMMOCK J, et al. Tandem mass spectrometric analysis of novel antineoplastic curcumin analogues [J]. In: Detection of Chemical, Biological, Radiological and Nuclear Agents for the Prevention of Terrorism. edn.: Springer, 2014: 223-231.

[112] ZANG R, LI D, TANG I C, et al. Cell-based assays in high-throughput screening for drug discovery [J]. International Journal of Biotechnology for Wellness Industries, 2012, 1 (1): 31-51.

[113] GUO C X, NG S R, KHOO S Y, et al. RGD-peptide functionalized graphene biomimetic live-cell sensor for real-time detection of nitric oxide molecules [J]. ACS Nano, 2012, 6 (8): 6944-6951.

[114] AGUSTIN Y E, TSAI S L. A high-throughput and selective method for the measurement of

surface areas of silver nanoparticles [J]. Analyst, 2015, 140 (8): 2618-2622.

[115] EDMONDSON R, BROGLIE J J, ADCOCK A F, et al. Three-dimensional cell culture systems and their applications in drug discovery and cell-based biosensors [J]. Assay Drug Dev Techn, 2014, 12 (4): 207-218.

[116] HOLGADO T M, QUINTANA M C, PINILLA J M. Electrochemical study of taxol (paclitaxel) by cathodic stripping voltammetry: Determination in human urine [J]. Microchem J, 2003, 74 (1): 99-104.

[117] TAEI M, HASSANPOUR F, SALAVATI H, et al. Highly selective electrochemical determination of taxol based on ds-DNA-modified pencil electrode [J]. Appl Biochem Biotechnol, 2015, 176 (2): 344-358.

[118] TAJIK S, TAHER M A, BEITOLLAHI H, et al. Electrochemical determination of the anticancer drug taxol at a ds-DNA modified pencil-graphite electrode and its application as a label-free electrochemical biosensor [J]. Talanta, 2015, 134: 60-64.

[119] HAHN Y, LEE H Y. Electrochemical behavior and square wave voltammetric determination of doxorubicin hydrochloride [J]. Archives of Pharmacal Research, 2004, 27 (1): 31-34.

[120] BAHNER N, REICH P, FRENSE D, et al. An aptamer-based biosensor for detection of doxorubicin by electrochemical impedance spectroscopy [J]. Anal Bioanal Chem, 2018, 410 (5): 1453-1462.

[121] GUO Y J, CHEN Y H, ZHAO Q, et al. Electrochemical sensor for ultrasensitive determination of doxorubicin and methotrexate based on cyclodextrin-graphene hybrid nanosheets [J]. Electroanalysis, 2011, 23 (10): 2400-2407.

[122] TING B P, ZHANG J, GAO Z, et al. A DNA biosensor based on the detection of doxorubicin-conjugated Ag nanoparticle labels using solid-state voltammetry [J]. Biosensors and Bioelectronics, 2009, 25 (2): 282-287.

[123] HATAMLUYI B, ES'HAGHI Z. A layer-by-layer sensing architecture based on dendrimer and ionic liquid supported reduced graphene oxide for simultaneous hollow-fiber solid phase microextraction and electrochemical determination of anti-cancer drug imatinib in biological samples [J]. Journal of Electroanalytical Chemistry, 2017, 801: 439-449.

[124] RODRÍGUEZ J, CASTAÑEDA G, LIZCANO I. Electrochemical sensor for leukemia drug imatinib determination in urine by adsorptive striping square wave voltammetry using modified screen-printed electrodes [J]. Electrochim Acta, 2018, 269: 668-675.

[125] CHEN H C, LUO K, LI K. A facile electrochemical sensor based on NiO-ZnO/MWCNT-COOH modified GCE for simultaneous quantification of imatinib and itraconazole [J]. J Electrochem Soc, 2019, 166 (8): B697-B707.

[126] TSENG T W, RAJAJI U, CHEN T W, et al. Sonochemical synthesis and fabrication of perovskite type calcium titanate interfacial nanostructure supported on graphene oxide sheets as a highly efficient electrocatalyst for electrochemical detection of chemotherapeutic drug [J]. Ultrasonics Sonochemistry, 2020, 69: 105242.

[127] DING L, HAO C, XUE Y, et al. A bio-inspired support of gold nanoparticles-chitosan nanocomposites gel for immobilization and electrochemical study of K562 leukemia cells [J]. Biomacromolecules, 2007, 8 (4): 1341-1346.

[128] KOVACIC P, SOMANATHAN R. Recent developments in the mechanism of anticancer agents based on electron transfer, reactive oxygen species and oxidative stress [J]. Anti-Cancer Agents in Medicinal Chemistry (Formerly Current Medicinal Chemistry-Anti-Cancer Agents), 2011, 11 (7): 658-668.

[129] KOVACIC P, SOMANATHAN R. Cell signaling and cancer: Integrated, fundamental approach involving electron transfer, reactive oxygen species, and antioxidants [J]. In: Cell Signaling & Molecular Targets in Cancer. edn.: Springer, 2012: 273-297.

[130] KOVACIC P, SOMANATHAN R. Nervous about developments in electron transfer-reactive oxygen species-oxidative stress mechanisms of neurotoxicity? [J]. In: Systems Biology of Free Radicals and Antioxidants. edn.: Springer, 2014: 1925-1944.

[131] CALAS-BLANCHARD C, CATANANTE G, NOGUER T. Electrochemical sensor and biosensor strategies for ROS/RNS detection in biological systems [J]. Electroanalysis, 2014, 26 (6): 1277-1286.

[132] DITTRICH P, IBANEZ A J. Analysis of metabolites in single cells-what is the best microplatform [J]. Electrophoresis, 2015, 36 (18): 2196-2206.

[133] EL-ALI J, SORGER P K, JENSEN K F. Cells on chips [J]. Nature, 2006, 442 (7101): 403-411.

[134] YEON J H, PARK J K. Cytotoxicity test based on electrochemical impedance measurement of HepG2 cultured in microfabricated cell chip [J]. Analytical Biochemistry, 2005, 341 (2): 308-315.

[135] ABIRI H, ABDOLAHAD M, GHAROONI M, et al. Monitoring the spreading stage of lung cells by silicon nanowire electrical cell impedance sensor for cancer detection purposes [J]. Biosensors and Bioelectronics, 2015, 68: 577-585.

[136] SZULCEK R, BOGAARD H J, VAN NIEUW AMERONGEN G P. Electric cell-substrate impedance sensing for the quantification of endothelial proliferation, barrier function, and motility [J]. Journal of visualized experiments: JoVE, 2014 (85): 51300.

[137] NERIMETLA R, KRISHNAN S. Electrocatalysis by subcellular liver fractions bound to carbon

nanostructures for stereoselective green drug metabolite synthesis [J]. Chemical Communications, 2015, 51 (58): 11681-11684.

[138] TRAN T B, NGUYEN P D, UM S H, et al. Real-time monitoring in vitro cellular cytotoxicity of silica nanotubes using electric cell-substrate impedance sensing (ECIS) [J]. Journal of Biomedical Nanotechnology, 2013, 9 (2): 286-290.

[139] ZHOU J, WU C, TU J, et al. Assessment of cadmium-induced hepatotoxicity and protective effects of zinc against it using an improved cell-based biosensor [J]. Sensors and Actuators a-Physical, 2013, 199: 156-164.

[140] HU N, ZHOU J, SU K, et al. An integrated label-free cell-based biosensor for simultaneously monitoring of cellular physiology multiparameter in vitro [J]. Biomedical microdevices, 2013, 15 (3): 473-480.

[141] MICHAELIS S, ROBELEK R, WEGENER J. Studying cell-surface interactions in vitro: A survey of experimental approaches and techniques [J]. In: Tissue Engineering Ⅲ: Cell-Surface Interactions for Tissue Culture. edn.: Springer, 2012: 33-66.

[142] GUO C X, NG S R, KHOO S Y, et al. RGD-peptide functionalized graphene biomimetic live-cell sensor for real-time detection of nitric oxide molecules [J]. ACS Nano, 2012, 6 (8): 6944-6951.

[143] LUO J, JIANG S S, ZHANG H Y, et al. A novel non-enzymatic glucose sensor based on Cu nanoparticle modified graphene sheets electrode [J]. Anal Chim Acta, 2012, 709: 47-53.

4 电化学生物传感器在UV诱导黑色素瘤细胞 $O_2^{·-}$ 的应用

4.1 引　言

　　紫外光（UV）是诱发人类皮肤病的重要影响因素之一，由于大气臭氧层的破坏，紫外线照射将不断增加最终导致人们遭受其伤害[1,2]。现在已经证明紫外线照射能够引起一些活性氧的产生[3-8]。活性氧引起的细胞氧化损伤被认为是一个关键的致癌因子[9,10]。在大多数情况下，来自太阳光的紫外线是由三个区域的波长组成。在高空大气中UVC被臭氧层所吸收，但UVB和UVA与大部分的皮肤疾病有关（见图4.1）。UVB的波长范围为290～320 nm，大部分由表皮和角化细胞的DNA吸收，而UVA（320～400 nm）能够深入到皮肤的真皮层，引起细胞的氧化应激[11,12]。由于UV光可以穿透生物体的不同皮肤层，通过不同的途径引起皮肤性疾病，UVA与UVB对皮肤性疾病的影响和作用机制已被报道[13-15]。例如，Petersen报道了UVA诱导产生的超氧阴离子可能与DNA的损伤有关[16]。UVB能够诱导活性氧的生成进而导致细胞的氧化损伤[17-19]。在许多活性氧自由基研究中超氧阴离子自由基是一个非常重要的自由基[16,20-22]。它在多种生物系统中作为分子氧的中间媒介物还原产生，它也可以由过渡金属离子氢氧自由基产生，例如 Fe^{2+} 和 Cu^{2+} [23]。此外，在动脉粥样硬化与神经系统变性疾病的发病机理中 $O_2^{·-}$ 与NO的反应能够导致高活性的过氧亚硝酸盐的形成[24-26]。它与羟基（—OH）结合后的产物会导致细胞DNA损坏，破坏人类机体功能[27]。因此，关于紫外线引起细胞释放 $O_2^{·-}$ 的研究具有重要意义，尤其是通过分析肿瘤细胞释放的 $O_2^{·-}$ 来评价一些未知的抗氧化复合物的抗氧化保护作用。

　　相关研究调查显示目前检测超氧阴离子的主要方法是基于探针标记实验[28,29]。细胞内荧光组织化学[30]、流式细胞术[19]和荧光光谱分析[31,32]是检测

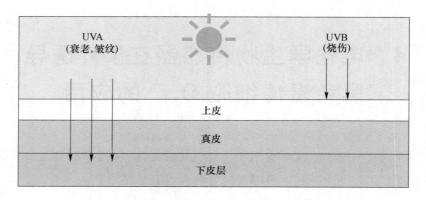

图 4.1　UV 对不同皮肤层的影响

$O_2^{\cdot-}$ 重要的方法，常用的荧光染料，例如，活性氧检测探针（DCFH-DA）[33,34]，二氢乙锭[35,36]和二氢罗丹明 123[37,38]等。通常检测细胞产生的 $O_2^{\cdot-}$ 可以采用细胞色素 C 还原法，在细胞悬液中加入细胞色素 C 通过光吸收值反映溶液中 $O_2^{\cdot-}$ 的生成量[39,40]。其他的检测方法如电子自旋共振法（ESR）被用于检测 UV 照射皮肤细胞产生的 $O_2^{\cdot-}$ [13,32]。但是这些探针标记方法需要昂贵的设备、复杂的测定程序、耗时且难以自动化，容易受其他物质的干扰。由于自由基的寿命短，因此自由基的检测就要求分析方法对目标物能够快速响应，以获得足够的信噪比[23,41]。电化学传感器由于具有设备简单、操作方便等优势，已成为实时监测自由基离子最具发展潜力的方法。Li 等人研制了钾掺杂多壁碳纳米管复合凝胶电化学生物传感器用于检测癌细胞释放的超氧阴离子自由基[23]。在前期工作中，我们制备了三明治型过氧化氢纸电极传感器，在三维环境中检测了细胞释放的过氧化氢[42]。这些研究表明，生物识别元件和电化学转导元件紧密结合制备电化学生物传感器，对于了解生物学过程具有很大的潜在应用价值。此外，由于检测使用样品的体积小，对于昂贵试剂使用也可实现，特别是罕见临床活检样本的保存和制作更具成本效益。

在这项研究中，我们设计了一个非标记的电化学传感器用于检测紫外光照射皮肤细胞产生的超氧阴离子水平。为了验证传感器的电化学分析能力，我们选用抗氧化药物模型（生育酚和圣草酚）研究角质化细胞和黑色素瘤细胞在紫外光照射下产生超氧阴离子水平的差异，分析非标记电化学生物传感器通过监测超氧阴离子水平筛选抗氧化药物的抗氧化能力。

4.2 结果与讨论

4.2.1 电化学检测装置

电化学检测装置：电化学检测装置如图 4.2 所示，电化学测试选用 CHI760E 电化学检测工作站（中国上海辰华仪器有限公司）；实验使用纳米功能化材料修饰玻碳电极，采用传统的三电极体系测试，玻碳电极作为工作电极，铂丝作为辅助电极，饱和甘汞电极作为参比电极，实验使用电极、电极抛光粉和抛光布等均购于中国上海辰华仪器有限公司；为了避免紫外光直接照射对电极的影响，电极从电解池的侧面插入并固定。另外，我们使用一片中间有空的铝箔纸覆盖在检测池的上面使紫外光只能照射在细胞上，避免紫外光照射到电极对检测结果的干扰。

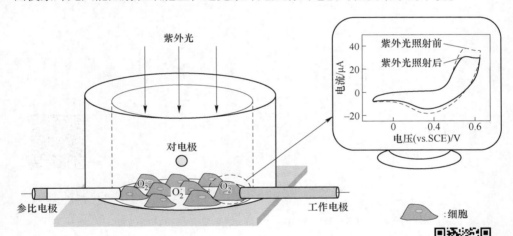

图 4.2　电化学传感器定量检测 UV 照射细胞产生的超氧阴离子

电化学检测平台是由纳米复合材料修饰的玻碳电极为工作电极，饱和甘汞电极为参比电极，铂丝为对电极的三电极体系组成，电化学工作站为 CHI760E。电化学传感器标定了不同浓度的超氧阴离子，检测装置平台如图 4.2 所示。

图 4.2 彩图

4.2.2　CNT/DNA@Mn$_3$(PO$_4$)$_2$ 纳米复合材料制备

超氧阴离子电化学传感器的制备：CNT/DNA@Mn$_3$(PO$_4$)$_2$ 纳米复合材料合成方法介绍如下[43]。首先，在烧杯中放入 9 mL 水、2.1 mg 的双链 DNA 和 1 mL、

0.1 mol/L MnSO₄ 在 60 ℃下旋转搅拌，然后再加入 1 mL、0.1 mol/L K₃PO₄ 和 9 mL 水继续搅拌直到溶液变透明即可。DNA@ Mn₃(PO₄)₂ 纳米混合物在 9000 r/min 下离心 10 min 收集沉淀，即获得 DNA@ Mn₃(PO₄)₂ 纳米复合材料（见图 4.3）。

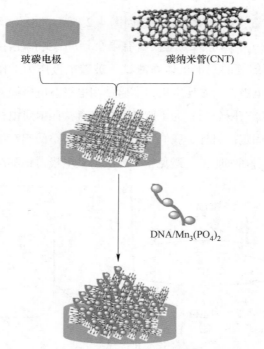

图 4.3　CNT/DNA@ Mn₃(PO₄)₂ 纳米复合材料合成过程

在实验室条件下，电化学方法中 $O_2^{\cdot-}$ 的产生方法之一[22,44]，参考下列公式：

$$KO_2 \Longrightarrow K^+ + O_2^{\cdot-} \tag{4.1}$$

超氧阴离子电化学传感器的制备：超氧阴离子电化学传感器的电极材料参考前期工作[43]。电极的制备过程简要概括如下：将预先准备的 0.5 mg/mL CNT 和 DNA@ Mn₃(PO₄)₂ 纳米复合材料滴加到电极表面室温晾干，通过此步骤即可完成电极的修饰。首先，利用循环伏安（CV）方法在 PBS 电解液中对 CNT/DNA@ Mn₃(PO₄)₂ 修饰的电极性能进行表征。通过在 0.1 mol/L PBS 中加入 KO₂ 以验证所制备的 $O_2^{\cdot-}$ 和 NO 传感器的性能和准确性。100 nmol/L 的 KO₂ 能够引起氧化峰电流值大约在 0.7 V 出现很强的增高（见图 4.4(a) 中的蓝线），再加入 SOD 后增加的峰电流值返回到基线水平（见图 4.4(a) 中的红线），由此可以说明电流值的增加是由于 $O_2^{·-}$ 的歧化反应引起的。为了进一步证明该传感器的检测灵敏度，使用时间电流曲线（i-t）对一系列 $O_2^{·-}$ 的浓度进行测定。电化学传感器在 5 s 内可以对 $O_2^{·-}$ 做出快速的响应，而且具有 10～200 nmol/L 的稳定检测范围。由超氧阴离子的校准曲线可获得传感器具有 2.5 nmol/L 的检测极限和 3.95 nA/(nmol·L^{-1}) 的敏感度（见图 4.4(b)）。通过时间-电流曲线检测电极对不同浓度的 $O_2^{·-}$ 的响应敏感性，发现 CNT/DNA@ $Mn_3(PO_4)_2$ 电化学传感器对 $O_2^{·-}$ 具有较好的灵敏度。综上所述，CNT/DNA@ $Mn_3(PO_4)_2$ 电化学传感器可用于检测 $O_2^{·-}$。

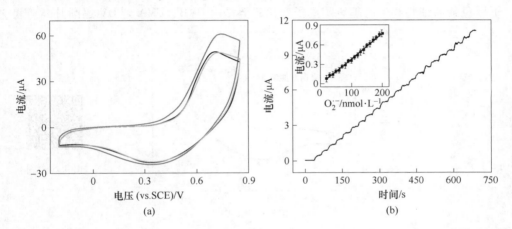

图 4.4　CNT/DNA@ $Mn_3(PO_4)_2$ 修饰电极检测 $O_2^{·-}$ （a）和不同浓度 KO_2 的时间电流曲线与浓度校正曲线（b）

（图（a）中循环伏安曲线检测 PBS 为黑色线，PBS 中加入 100 nmol/L KO_2 为蓝色线，PBS 中加入 100 nmol/L KO_2 和 SOD 为红色线）

图 4.4 彩图

4.2.3　超氧阴离子电化学传感器量化紫外光诱导黑色素瘤细胞产生的 $O_2^{·-}$

通过上述实验对 CNT/DNA@ $Mn_3(PO_4)_2$ 电化学传感器的制备和性能表征进行评价，然后选用电化学方法检测不同波长紫外光照射诱导的 HaCaT 和 A375 细胞产生的超氧阴离子，并对电化学传感器的可行性进行评价。由实验可知，CNT/DNA@ $Mn_3(PO_4)_2$ 电化学传感器可以检测体系中存在的超氧阴离子。

原位检测由 UV 照射细胞产生的超氧阴离子：首先，将几个自制的直径为 1 cm 的细胞培养盘放到直径为 3.5 cm 的培养皿中。然后，将 HaCaT 和 A375 细胞按照 $5×10^5$/mL 的密度接种在培养皿中孵育 8 h，等细胞贴壁后再加入生育酚

(25 μmol/L)和圣草酚（50 μmol/L）继续孵育 12 h。最后，将贴附有细胞的培养盘转移预先加入 2.5 mL 的无血清培养液的电化学检测室（见图 4.2）中。电化学检测工作站记录在紫外光照射下细胞产生的超氧阴离子。未经抗氧化药物生育酚和圣草酚处理的细胞作为实验的对照组，对照组细胞释放的超氧阴离子同样采用电化学工作站采集电信号数据。

超氧阴离子电化学传感器的校准曲线可以用于量化 UVR 诱导细胞产生的超氧阴离子。图 4.5(a) 展示了 HaCaT 细胞分别暴露于 UV(105 kJ/m^2)、UVA(100 kJ/m^2)和 UVB(5 kJ/m^2)下电化学循环伏安曲线检测不同波长紫外光照射下黑色素瘤细胞产生的超氧阴离子变化趋势是由 UVB、UVA 到 UV 逐渐升高。表

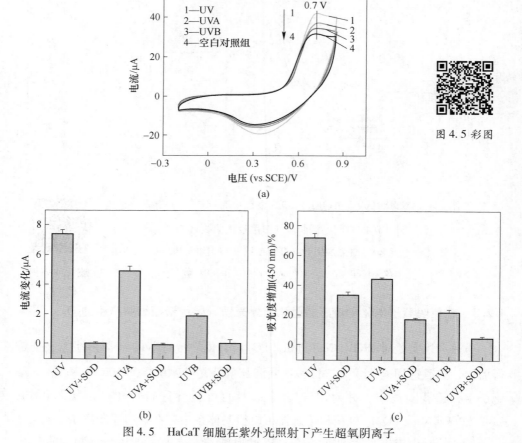

图 4.5 HaCaT 细胞在紫外光照射下产生超氧阴离子

（a）HaCaT 细胞产生的超氧阴离子的循环伏安曲线；（b）在紫外光照射下与对照组相比峰电流变化的柱形图；（c）用酶标仪 450 nm 处光吸收变化表示超氧化物检测试剂盒检测 UV 照射下超氧阴离子的产生量

明不同剂量的紫外光诱导细胞释放的 $O_2^{\cdot-}$ 水平不同。图 4.5(b) 是由不同波长的紫外光触发细胞产生峰电流变化的三次独立实验数据做出的柱状图。此图展现出当与 SOD 共同孵育细胞时循环伏安曲线的峰电流值降低，与对照组相比，表明电化学检测信号强弱直接反映出紫外线照射细胞产生的超氧阴离子的量的多少。此外，为了验证电化学方法检测结果的灵敏度，选用超氧化物检测试剂盒检测紫外光照射 HaCaT 细胞产生的超氧阴离子的水平（见图 4.5(c)），当使用超氧化物歧化酶预处理细胞后，在相同紫外光处理下产生的 $O_2^{\cdot-}$ 则没有增加，说明 SOD 可以有效清除细胞产生的 $O_2^{\cdot-}$。我们观察到试剂盒检测结果与电化学获得结果一致。从图 4.6 中可以看出 A375 细胞在紫外光照射下产生的 $O_2^{\cdot-}$ 与图 4.5 中 HaCaT

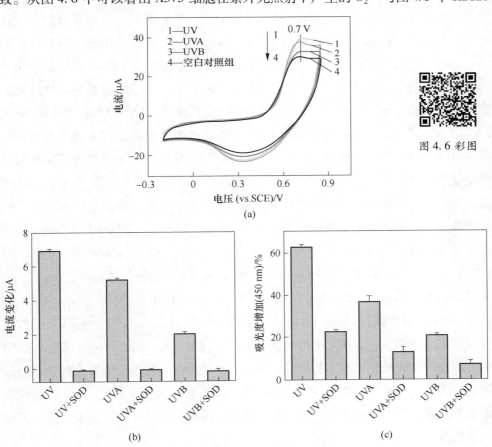

图 4.6 A375 细胞在紫外光照射下产生超氧阴离子

(a) A375 细胞产生的超氧阴离子的循环伏安曲线；(b) 在紫外光照射下与对照组相比峰电流变化的柱形图；
(c) 用酶标仪 450 nm 处光吸收变化表示超氧化物检测试剂盒检测 UV 照射下超氧阴离子的产生量

细胞趋势是相同的，不同的是不同细胞抵御紫外光的能力存在差异。选用电化学传感器检测紫外光辐射引起的 HaCaT 和 A375 细胞产生的 $O_2^{·-}$ 水平与传统的荧光染色得到的检测结果一致，而且可以排除细胞内常见干扰物的干扰。电化学传感器具有样品消耗少，实验操作时间短，对细胞无损伤等优点。

4.2.4 HaCaT 和 A375 细胞在紫外光照射下的生长能力

细胞在 UV 照射下由于发生氧化损伤引起凋亡。生育酚和圣草酚是两种具有抗氧化作用的抗氧化剂，对细胞的氧化损伤有一定的保护作用。我们通过 HaCaT 和 A375 细胞研究了抗氧化药物预处理后细胞生长能力的变化，寻找出这两种抗氧化药物的有效作用浓度。使用 MTT 测定生育酚和圣草酚对紫外光照射角质化细胞（HaCaT）和黑色素瘤细胞（A375）的保护作用进行了评价。简言之：第一，将细胞按每孔 1×10^4/mL 细胞密度养在 96 孔板中，分别用生育酚和不用生育酚孵育 12 h 后，圣草酚实验组与生育酚相同；第二，将细胞暴露于紫外光下照射，继续培养 24 h；第三，在每孔中加入 MTT 溶液（0.5 mg/mL）10 μL 并在 37 ℃下孵育 2 h 后，在各孔中加入 100 μL 由 2-丙醇、聚乙二醇辛基苯基醚和盐酸制备的细胞裂解缓冲液并溶解，此时可以观察到活细胞线粒体中的琥珀酸脱氢酶能使外源性 MTT 还原为水不溶性的蓝紫色结晶甲瓒并沉积在细胞中，而死细胞无此功能；第四，通过酶标仪在 570 nm 处测定吸光度，通过照射组与对照组细胞成活率百分比来表示实验结果。图 4.7(a) 和（b）表明 HaCaT 细胞在 UV 照射下生育酚（0 μmol/L，10 μmol/L，25 μmol/L 和 50 μmol/L）和圣草酚（0 μmol/L，50 μmol/L，100 μmol/L 和 200 μmol/L）细胞的成活率，图 4.7(c) 和（d）显示了 A375 细胞选用生育酚（0 μmol/L，10 μmol/L，25 μmol/L 和 50 μmol/L）和圣草酚（0 μmol/L，25 μmol/L，50 μmol/L 和 75 μmol/L）药物浓度时细胞成活率。从图 4.7 中可知生育酚（25 μmol/L）和圣草酚（50 μmol/L）孵育的细胞在紫外光照射下其成活率明显高于其他组，细胞成活率大约为半数致死量。因此，可推出生育酚（25 μmol/L）和圣草酚（50 μmol/L）的药物浓度对紫外光诱导的细胞氧化损伤可以起到很好的保护作用，可以作为本工作的最佳药物浓度。

4.2.5 生育酚可以减少紫外线诱导黑色素瘤细胞产生的 $O_2^{·-}$

为了证明电化学传感器作为评价化合物的抗氧化能力可行性的分析工具，我

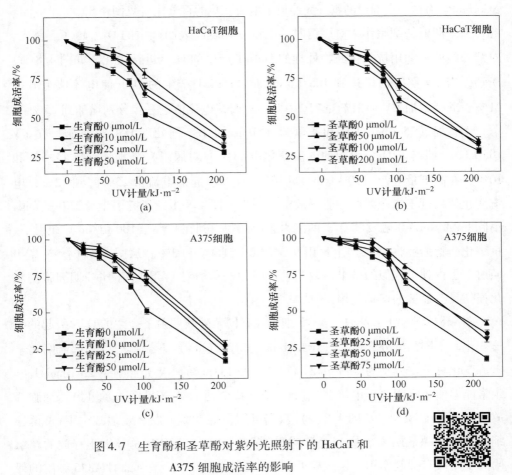

图 4.7 生育酚和圣草酚对紫外光照射下的 HaCaT 和
A375 细胞成活率的影响

（a）不同浓度的生育酚对 UV 照射下 HaCaT 细胞生长的影响；（b）不同浓度的圣草酚对 UV 照射下 HaCaT 细胞生长的影响；（c）不同浓度的生育酚对 UV 照射下 A375 细胞生长的影响；（d）不同浓度的圣草酚对 UV 照射下 A375 细胞生长的影响

们选用生育酚和圣草酚预处理细胞，然后进行 UVR 照射探讨化合物抗氧化能力。在这项研究中，电化学传感器直接用于测量在紫外线照射下预先用抗氧化剂处理细胞产生的 $O_2^{\cdot-}$，并进一步研究了生育酚和圣草酚对细胞在 UVA、UVB 和 UV 的诱导下产生的自由基的清除作用。为此，使用未照射的细胞作为实验对照组，计算 0.7 V 电位处的电流变化率公式：

$$\Delta \text{Current}\% = [(C_i - C_0)/C_0] \times 100\% \tag{4.2}$$

式中，C_i 为在紫外光照射下测得的电流值；C_0 为未照射紫外光条件下测得的电流值；

ΔCurrent%为在一定剂量的紫外光照射下电流变化的百分比（见图4.8）。

当未照射的细胞作为对照组时，电流变化越高表明产生的$O_2^{·-}$越多。首先，对紫外线照射和用抗氧化药物处理的HaCaT产生的$O_2^{·-}$进行量化。如图4.8(a)所示，当UV照射剂量达到105 kJ/m^{2}时，HaCaT产生的$O_2^{·-}$使电流值上升了21%。继续增加UV照射剂量到210 kJ/m^{2}时，电流值进一步分别增加到25%。

生育酚（25 μmol/L）和圣草酚（50 μmol/L）预处理HaCaT细胞，在UV 105 kJ/m^{2}照射下$O_2^{·-}$引起的电流变化由21%（对照组）分别降到了11.5%和10%；在高剂量UV（210 kJ/m^2）照射下$O_2^{·-}$引起的电流由25%降到12%，电流变化值接近于未照射组细胞释放量的一半。图4.8(b)研究了生育酚和圣草酚消除UVA和UVB触发的$O_2^{·-}$的能力，由图观察到在UVA（100 kJ/m^2）照射下，抗氧化药物能够将$O_2^{·-}$产生量由9.5%减少到4%和3%。从图4.8(c)中UVB照射下也得到了相似的结果，说明生育酚和圣草酚对UVR照射HaCaT细胞的氧化损伤具有显著的保护作用。

图4.8(d)~(f)表示了A375细胞在不同紫外线照射下产生的$O_2^{·-}$引起的电流变化。不同的紫外光照射（105 kJ/m^2，210 kJ/m^2）下细胞产生的超氧阴离子电流变化百分比分别为17%和22%。在105 kJ/m^{2}的UV光照射下25 μmol/L维生素能够使超氧阴离子电流变化值由25%减少到11%（见图4.8(d)），细胞释放的$O_2^{·-}$明显减少。从图4.8(e)数据可知，生育酚可以有效地减少UVA诱导细胞释放的超氧阴离子从11.5%减少到2%，而对于UVB诱导细胞释放的超氧阴离子则没有抑制作用，仅减少了4%（见图4.8(f)）。圣草酚对A375细胞的保护作用与生育酚不同，圣草酚可以有效地减少UVB诱导细胞释放的$O_2^{·-}$（从11.7%降低到1.5%）。从图4.8中说明生育酚对UVA诱导的细胞氧化损伤具有有效的抗氧化保护作用，然而对于UVB诱导的氧化损伤则没有保护作用，圣草酚的抗氧化作用与生育酚相比则相反。

电化学传感器除了可以用于量化紫外光照射时细胞产生的$O_2^{·-}$，还可以监测在紫外线照射后期细胞产生$O_2^{·-}$水平的变化情况，这可以进一步评价抗氧化剂生育酚和圣草酚的抗氧化能力。图4.9(a)~(c)展示了HaCaT细胞在紫外光诱导后产生的$O_2^{·-}$引起的电流变化。当细胞不用抗氧化药物处理时，随着UV剂量的增加产生的$O_2^{·-}$浓度逐渐增加。首先给予UV（105 kJ/m^2）照射HaCaT细胞，照射后0 min和60 min时$O_2^{·-}$的峰电流值分别增加了22%和13%。相同条件下，

4.2 结果与讨论

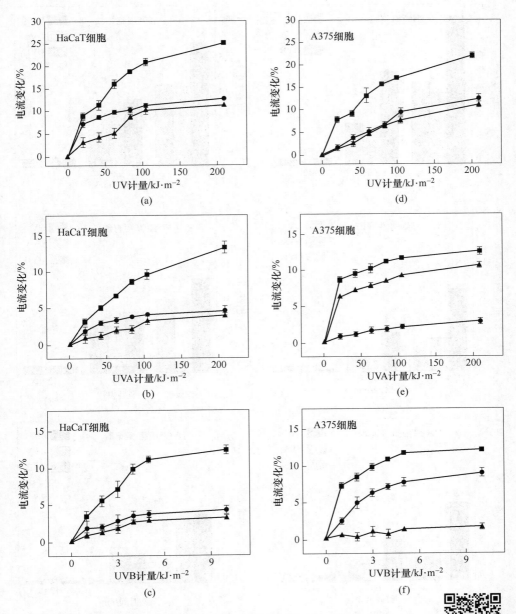

图 4.8 电化学方法量化 HaCaT 和 A375 细胞释放的 $O_2^{\cdot -}$

(黑色线为对照组;蓝色线为生育酚预处理细胞;红色线为圣草酚预处理细胞 ($n=3$))

(a) ~ (c) HaCaT 分别在 UV、UVA 和 UVB 照射下 $O_2^{\cdot -}$ 引起的电流值变化;

(d) ~ (f) A375 分别在 UV、UVA 和 UVB 照射下 $O_2^{\cdot -}$ 引起的电流值变化

图 4.8 彩图

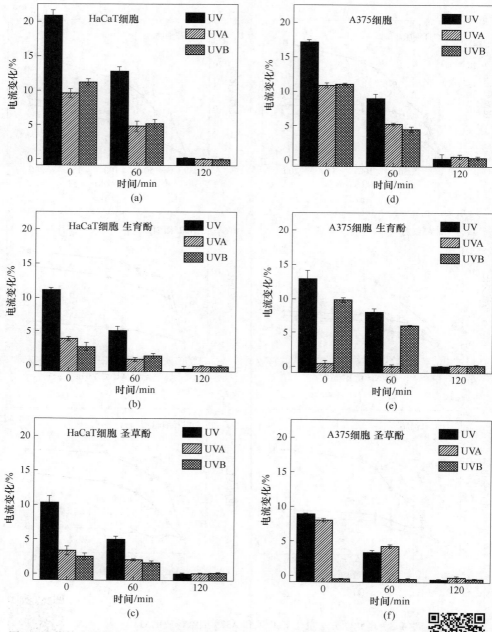

图 4.9 紫外光照射之后电化学监测 HaCaT 和 A375 细胞释放的 $O_2^{\cdot-}$ 消解情况
（a）~（c）抗氧化药物预处理的 HaCaT 分别在 UV、UVA 和 UVB 照射后 $O_2^{\cdot-}$ 的变化；
（d）~（f）抗氧化药物预处理的 A375 分别在 UV、UVA 和 UVB 照射后 $O_2^{\cdot-}$ 的变化（$n=3$）

图 4.9 彩图

在 UVA 和 UVB 照射下，产生的 $O_2^{·-}$ 与 UV 相比较少。两种抗氧化药物处理细胞后消除了 UV 照射产生的 $O_2^{·-}$（见图 4.9（b）和（c））。当用抗氧化药物处理 A375 细胞后，我们观察到类似的现象（见图 4.9）。当 UV 刺激细胞后 120 min 产生的 $O_2^{·-}$ 消退到了对照组水平。我们观察到一个有趣的现象，用抗氧化剂生育酚预处理的黑色素瘤 A375 细胞在 UVA 紫外光辐照后期与对照组相比产生的超氧阴离子水平是稳定的（见图 4.9（e）），但是对于用生育酚预处理的黑色素瘤 A375 细胞在 UVB 照射后 120 min 超氧阴离子消退到了对照组水平（见图 4.9（f））。相反，圣草酚能够轻微的抑制 UVA 刺激产生的 $O_2^{·-}$，但是可以显著阻止照射后 UVB 产生 $O_2^{·-}$。

通过分析图 4.8 和图 4.9 中电化学检测结果，我们发现：（1）UVA 和 UVB 触发 HaCaT 和 A375 细胞产生 $O_2^{·-}$ 的能力不同。UV 诱导细胞产生 $O_2^{·-}$ 的能力不能简单地视为是 UVA 和 UVB 的效果总和，例如，通过 UV（0，21 kJ/m²，42 kJ/m²，63 kJ/m²，84 kJ/m²，105 kJ/m²，210 kJ/m²）刺激从细胞中产生 $O_2^{·-}$ 的量低于 UVA（0，20 kJ/m²，40 kJ/m²，60 kJ/m²，80 kJ/m²，100 kJ/m²，200 kJ/m²）和 UVB（0，1 kJ/m²，2 kJ/m²，3 kJ/m²，4 kJ/m²，5 kJ/m²，10 kJ/m²）的总量。（2）生育酚和圣草酚在减少紫外线诱导产生的 $O_2^{·-}$ 中具有不同的抗氧化作用。生育酚和圣草酚能够很好地抑制 UVA 和 UVB 诱导的 HaCaT 细胞产生 $O_2^{·-}$，然而对于 A375 细胞，生育酚和圣草酚则分别对 UVA 和 UVB 诱导的损伤有保护作用。

4.2.6 电化学传感器预测生育酚在细胞成活能力中的保护作用

为进一步分析生育酚在细胞存活中的有效作用，我们选用 HaCaT 和 A375 细胞在不同剂量 UV、UVA 和 UVB 照射下进行，研究细胞在生育酚和圣草酚预处理细胞与对照组细胞的生存能力。这两种抗氧化药物的有效抗氧化能力通过细胞存活率公式进行评估，如下所示：

$$细胞存活\% = [(紫外光照射组细胞 - 对照组)/对照组] \times 100\% \quad (4.3)$$

A_T 为紫外光照射组细胞的吸光度值，A_c 为对照组（未照射紫外光组）细胞的吸光度值。通过上述公式的计算，在 UV 照射下细胞的成活率是以对照组（未照射紫外光组）为标准计算的。未用抗氧化药物处理的 HaCaT 用 UV 105 kJ/m² 剂量照射下，其成活率迅速降低了 50%，生育酚和圣草酚预处理的细胞可以显著地提高细胞的成活率，分别为 82% 和 76.5%（见图 4.10（a））。HaCaT 细胞遭受 UVA 损伤后细胞的成活率可以通过使用抗氧化药物得以挽救（见图 4.10

(b)),对于 UVB 诱导的 HaCaT 细胞损伤也可以得到挽救(见图 4.10(c))。图 4.10(d)~(f) 研究了两种抗氧化药物对黑色素瘤 A375 细胞的生存能力的保护

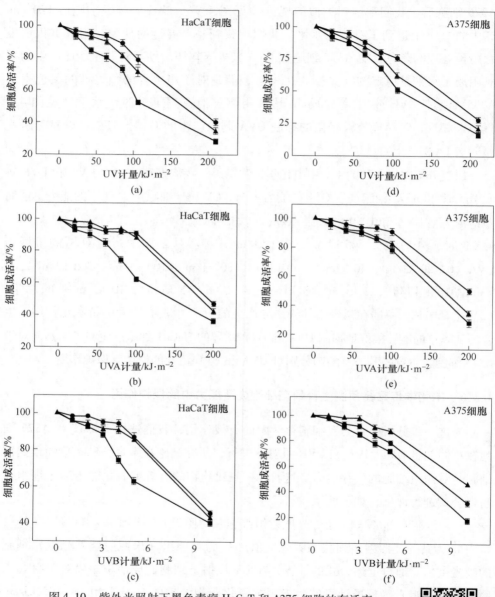

图 4.10 紫外光照射下黑色素瘤 HaCaT 和 A375 细胞的存活率
(未照射组为黑色线,生育酚预处理细胞为蓝色线,圣草酚预处理细胞为红色线)
(a)~(c) HaCaT 细胞在 UV、UVA 和 UVB 照射下细胞的成活率;
(d)~(f) A375 细胞在 UV、UVA 和 UVB 照射下细胞的成活率($n=3$)

图 4.10 彩图

作用。当细胞暴露于 UV 剂量（105 kJ/m²）时，如图 4.10(e) 所示，通过研究 UVA 诱导的细胞损伤时发现生育酚与圣草酚相比具有更好的抗氧化能力，生育酚与圣草酚处理后细胞的存活率分别为 90% 和 80%。相反，UVB 诱导的损伤能够通过圣草酚有效的抑制，但生育酚则没有此作用。

MTT 细胞成活率实验表明：（1）生育酚与圣草酚能够保护 UV 照射的 HaCaT 细胞免受损伤；（2）对于 A375 细胞，生育酚能够消除 UVA 诱导的氧化损伤，而圣草酚可以消除 UVB 诱导的氧化损伤。重要的是 MTT 实验结果与电化学传感器检测结果是一致的。抗氧化药物对紫外光诱导的损伤表现出不同的保护作用。综上所述，使用 MTT 细胞成活率实验，我们发现生育酚对 UVA 诱导的细胞氧化损伤具有一定的保护作用。重要的是，从 MTT 细胞成活率实验获得数据与电化学传感器检测的 $O_2^{\cdot-}$ 的结果（见图 4.7）是一致的。

4.2.7 对紫外线诱导细胞氧化损伤时不同检测方法比较

在这项研究中，MTT 实验可以评估由于紫外线诱导的细胞生存能力的降低，它也可以监测一定时间内细胞的生存能力，进而可以展示出长时间的外界刺激对细胞生存能力的影响[43,45]。近 40 年大量的研究工作都是通过荧光探针标记法描述了细胞内活性氧的变化。但是，荧光探针标记法也表现出对于氧化剂瞬时通量的检测面临着巨大的困难[46,47]。最重要的是，荧光探针标记法在用于原位监测细胞在一段时间内的刺激下细胞内部的变化响应时面临着巨大的挑战。非标记电化学传感器的优点在一些研究工作中已被详细介绍[42,48,49]，在本研究中，对不同的研究方法进行列表阐述，见表 4.1。

表 4.1 UV 诱导氧化损伤的不同检测方法比较

项　目	MTT 实验	荧光实验	电化学检测
标记	是	是	否
实时	否	否	是
样品消耗	大量	大量	少量
实验时间	24~72 h	1~2 h	几分钟内
总体比较	细胞活性催化酶来间接评价细胞的生长，评估处理细胞的整体存活能力	荧光探针标记目标分子不能够实时检测染色过程影响细胞的生长状况	原位检测对细胞没有损伤，检测快速，可用于高通量筛选，实验仪器费用低

为了验证电化学传感器作为一种分析工具用于复合物抗氧化能力检测的可行性，我们选用传统的荧光标记法对细胞内活性氧进行检测，具体实验是使用紫外光诱导细胞氧化损伤，再应用标准的活性氧检测试剂盒（一种利用荧光探针DCFH-DA进行活性氧检测的试剂盒）检测细胞内活性氧生成量。实验简要步骤如下：收集 HaCaT 和 A375 悬浮细胞，约按每孔 1×10^6 个细胞把细胞养在 24 孔板中，在 5% 二氧化碳和 37 ℃ 的细胞培养箱中孵育 6 h。使用活性氧检测试剂盒（DCFH-DA）对细胞进行细胞内的活性氧荧光染色。活性氧荧光探针以 1:1000 的比例进行稀释后加入 24 孔板中 37 ℃ 下避光孵育 20 min，然后将多余的 DCFH-DA 荧光染料用 PBS（1.0 mmol/L）清洗三次，使用荧光倒置显微镜对细胞进行观察并采集图片。从实验结果可以观察到紫外光确实可以诱导绿色荧光信号的增强，出现绿色荧光是因为此荧光染料使用 488 nm 激发波长，525 nm 发射波长引起的。图 4.11 展示了通过三次独立的荧光染色实验对 HaCaT 细胞内活性氧的荧光变化进行了量化。绿色荧光信号随着 UV 剂量的增加而逐渐升高，并且当细胞暴露在 210 kJ/m² UV 照射时荧光的信号强度迅速增加。在这个荧光染色实验研究中我们观察到细胞内活性氧的升高是随着 UV 剂量的变化而变化的，这与相关的研究报告[20]以及电化学传感器的检测结果是一致的。进一步增强 UV 剂量（210 kJ/m²）仅能使活性氧的荧光信号强度缓慢增加。

我们特别研究了抗氧化物在 UV 剂量 105 kJ/m²（UVA 100 kJ/m² + UVB 5 kJ/m²）时的有效保护作用。超氧化物检测试剂盒被用来量化 $O_2^{·-}$ 诱导的检测工作液比色度变化，然后通过多功能酶标仪检测吸光度值，进而评价超氧阴离子的水平。超氧化物检测试剂盒是一种用于超氧化物快速高灵敏度检测的试剂盒，试剂盒是利用超氧化物可以还原 WST-1 产生可溶性的有色物质为基础来检测超氧化物。具体操作步骤如下：第一，细胞约按每孔 1×10^5/mL 密度养在 96 孔板中，用抗氧化复合物和不用抗氧化复合物孵育细胞 24 h。第二，吸除培养液并用 PBS（1.0 mmol/L）清洗一次。第三，每孔加入 200 μL 超氧化物检测工作液，37 ℃ 下孵育 3 min。在待测样品孔中加入酵母多糖诱导细胞产生超氧化物，刺激 60 min。空白对照组不加入酵母多糖刺激物。阳性对照组孔中加入 2 μL 超氧化物歧化酶用以验证整个检测体系的可行性。第四，将细胞暴露于 UV、UVA 和 UVB 下 10 min 后，然后在酶标仪 450 nm 处测定吸光度值，使用大于 600 nm 的波长作为参考波长进行双波长测定。实验数据通过样品组与对照组相比较吸光度值升高的百分比表示。从实验结果可知抗氧化物能够减少 HaCaT 和 A375 细胞在

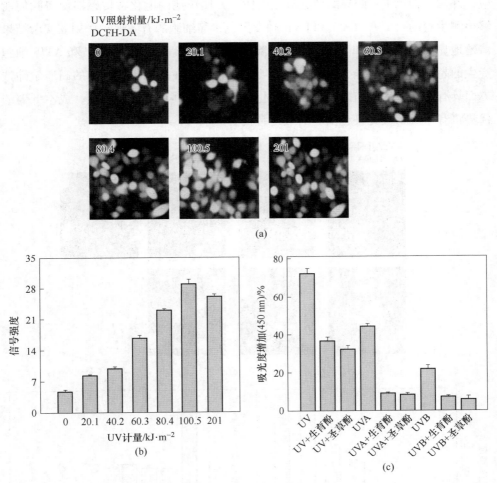

图 4.11 荧光染色法描述 UV 照射 HaCaT 细胞引起的
细胞内活性氧的产生

(a) DCFH-DA 染色的荧光显微镜图片；(b) 通过 Image J 量化的 DCFH-DA
染色荧光强度的柱状图，$n=3$；(c) 超氧化物检测试剂盒量化从
HaCaT 细胞中产生的超氧阴离子，$n=3$

图 4.11 彩图

UV 照射下产生的超氧阴离子。抗氧化剂具有不同的抗氧化功效对于不同波长的紫外线诱导的 A375 细胞氧化损伤的保护作用也是不同的（见图 4.12）。标准的分子标记技术与电化学方法都可以用于量化紫外线照射和抗氧化剂的抗氧化作用，展示了很好的相关性，此方法对于细胞生物学的深入研究具有促进作用。此外，$CNT/DNA@Mn_3(PO_4)_2$ 电化学传感器可以展示出超氧阴离子的氧化还原反

应,并用于检测 UV 诱导的细胞氧化损伤。方便快捷的电化学传感器使得我们能够检测生育酚对 UV、UVA 和 UVB 触发黑色素瘤细胞的有效保护作用。实验结果清楚地显示出生育酚和圣草酚可以消除 UVA 和 UVB 照射下 HaCaT 和 A375 细胞产生的超氧阴离子的能力不同,这表明电化学方法在抗氧化化合物的高通量筛选方面具有潜在应用价值。我们相信电化学方法由于其具有操作便捷,成本低廉的优势将有望成为荧光染色实验的最终替代者。

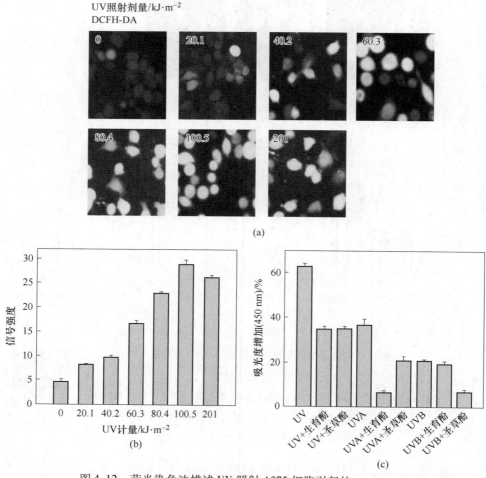

图 4.12　荧光染色法描述 UV 照射 A375 细胞引起的细胞内活性氧的产生

(a) DCFH-DA 染色的荧光显微镜图片;(b) 通过 Image J 量化的 DCFH-DA 染色荧光强度的柱状图,$n=3$;(c) 超氧化物检测试剂盒量化从 A375 细胞中产生的超氧阴离子,$n=3$

图 4.12 彩图

4.3 小　　结

紫外线照射损伤人体的免疫系统刺激皮肤细胞产生的自由基与皮肤癌有关。超氧阴离子（$O_2^{\cdot-}$）是一种重要的活性氧自由基，它参与紫外光照射诱导的皮肤细胞氧化损伤。在这个工作中，我们制备了单壁碳纳米管、核糖核苷酸和磷酸锰纳米复合材料生物传感器，实现了在不同的紫外光照射下对黑色素瘤细胞产生的 $O_2^{\cdot-}$ 浓度的检测。通过电化学传感器检测抗氧化药物生育酚和圣草酚处理皮肤细胞后 $O_2^{\cdot-}$ 的释放水平，验证抗氧化药物的有效抗氧化作用。电化学传感器通过电流值的变化对 $O_2^{\cdot-}$ 浓度进行量化，实验结果表明：对于黑色素瘤细胞，生育酚能够有效清除 UVA 照射产生的超氧阴离子，而对于 UVB 照射产生的 $O_2^{\cdot-}$ 具有较弱的清除能力；然而，圣草酚抗氧化作用情况与维生素相反，圣草酚可以有效地清除 UVB 照射产生的 $O_2^{\cdot-}$，对于 UVA 照射产生的 $O_2^{\cdot-}$ 具有较弱的清除能力。因此，非标记电化学传感器可以检测自由基并有望用于抗氧化药物的筛选。

实验结果表明：生育酚和圣草酚对于紫外线照射下的 HaCaT 细胞具有相似的作用。但是，对于 A375 细胞生育酚能够有效地抑制 UVA 的超氧阴离子自由基，圣草酚对 UVB 触发的超氧阴离子更有效。电化学传感器在紫外光生物学研究中的应用具有重要作用，因此电化学传感器可以作为量化 $O_2^{\cdot-}$ 的一种有效工具被用来解释抗氧化药物对细胞的生长具有不同的保护作用的机制。非标记电化学传感器与细胞内的荧光染色法检测结果一致，电化学传感器方法在紫外光照射下抗氧化药物的筛选具有潜在的应用价值，并且有助于研究抗氧化药物的作用机制。在这项研究中，我们设计的非标记的电化学传感器可以检测紫外光照射皮肤细胞产生的超氧阴离子水平。为了验证传感器的电化学分析能力，我们选用抗氧化药物模型（生育酚和圣草酚）消除黑色素瘤细胞在紫外线照射下产生的超氧阴离子。通过研究得出如下结论：（1）黑色素瘤 A375 细胞暴露于紫外线辐射 105 kJ/cm^2 时下产生的 $O_2^{\cdot-}$ 迅速增加，随后细胞出现细胞轮廓皱缩进而凋亡形态；（2）分别用 25 μmol/L 的生育酚和 50 μmol/L 圣草酚预处理细胞，然后选用同样剂量的紫外线照射细胞发现产生的 $O_2^{\cdot-}$ 水平下降；（3）不同波长的紫外光对细胞的氧化损伤程度不同，与 UV 和 UVB 相比，生育酚对 UVA 诱导的氧化损伤有较明显的保护作用；（4）将电化学测量数据与细胞存活率实验、常规的

ROS荧光染色和超氧化物试剂盒进行比较可知,我们所构建的非标记电化学方法有望用于抗氧化药物的高通量筛选。

参 考 文 献

[1] SIMON M M, REIKERSTORFER A, SCHWARZ A, et al. Heat-shock protein-70 overexpression affects the response to ultraviolet-light in murine fibroblasts-evidence for increased cell viability and suppression of cytokine release [J]. J Clin Invest, 1995, 95 (3): 926-933.

[2] GODAR D E. UV doses worldwide [J]. Photochem Photobiol, 2005, 81 (4): 736-749.

[3] LIEBEL F, KAUR S, RUVOLO E, et al. Irradiation of skin with visible light induces reactive oxygen species and matrix-degrading enzymes [J]. J Invest Dermatol, 2012, 132 (7): 1901-1907.

[4] RASS K, REICHRATH J. UV damage and DNA repair in malignant melanoma and nonmelanoma skin cancer [J]. Adv Exp Med Biol, 2008, 624: 162-178.

[5] DE GRUIJL F R, VAN KRANEN H J, MULLENDERS L H F. UV-induced DNA damage, repair, mutations and oncogenic pathways in skin cancer [J]. J Photoch Photobio B, 2001, 63 (1/2/3): 19-27.

[6] KULMS D, ZEISE E, POPPELMANN B, et al. DNA damage, death receptor activation and reactive oxygen species contribute to ultraviolet radiation-induced apoptosis in an essential and independent way [J]. Oncogene, 2002, 21 (38): 5844-5851.

[7] ARMSTRONG B K, KRICKER A. The epidemiology of UV induced skin cancer [J]. J Photoch Photobio B, 2001, 63 (1/2/3): 8-18.

[8] DIZDAROGLU M, JARUGA P, BIRINCIOGLU M, et al. Free radical-induced damage to DNA: Mechanisms and measurement [J]. Free Radical Biology and Medicine, 2002, 32 (11): 1102-1115.

[9] ZHANG Y G, MATTJUS P, SCHMID P C, et al. Involvement of the acid sphingomyelinase pathway in UVA-induced apoptosis [J]. Journal of Biological Chemistry, 2001, 276 (15): 11775-11782.

[10] BERWICK M, LACHIEWICZ A, PESTAK C, et al. Solar UV exposure and mortality from skin tumors [J]. Adv Exp Med Biol, 2008, 624: 117-124.

[11] TYRRELL R M, PIDOUX M. Singlet oxygen involvement in the inactivation of cultured human-fibroblasts by uva (334 nm, 365 nm) and near-visible (405 nm) radiations [J]. Photochem Photobiol, 1989, 49 (4): 407-412.

[12] DUTHIE M S, KIMBER I, DEARMAN R J, et al. Differential effects of UVA1 and UVB radiation on langerhans cell migration in mice [J]. J Photoch Photobio B, 2000, 57 (2/3): 123-131.

[13] HERRLING T, JUNG K, FUCHS J. Measurements of UV-generated free radicals/reactive oxygen species (ROS) in skin [J]. Spectrochim Acta A, 2006, 63 (4): 840-845.

[14] KRUTMANN J. The interaction of UVA and UVB wavebands with particular emphasis on signalling [J]. Prog Biophys Mol Bio, 2006, 92 (1): 105-107.

[15] SYED D N, AFAQ F, MUKHTAR H. Differential activation of signaling pathways by UVA and UVB radiation in normal human epidermal keratinocytes [J]. Photochem Photobiol, 2012, 88 (5): 1184-1190.

[16] PETERSEN A B, GNIADECKI R, VICANOVA J, et al. Hydrogen peroxide is responsible for UVA-induced DNA damage measured by alkaline comet assay in HaCaT keratinocytes [J]. J Photoch Photobio B, 2000, 59 (1/2/3): 123-131.

[17] WOLFLE U, ESSER P R, SIMON-HAARHAUS B, et al. UVB-induced DNA damage, generation of reactive oxygen species, and inflammation are effectively attenuated by the flavonoid luteolin in vitro and in vivo [J]. Free Radical Biology and Medicine, 2011, 50 (9): 1081-1093.

[18] YASUL H, HAKOZAKI T, DATE A, et al. Real-time chemiluminescent imaging and detection of reactive oxygen species generated in the UVB-exposed human skin equivalent model [J]. Biochem Bioph Res Co, 2006, 347 (1): 83-88.

[19] PEUS D, VASA R A, MEVES A, et al. H_2O_2 is an important mediator of UVB-induced EGF-receptor phosphorylation in cultured keratinocytes [J]. J Invest Dermatol, 1998, 110 (6): 966-971.

[20] REZVANI H R, MAZURIER F, CARIO-ANDRE M, et al. Protective effects of catalase overexpression on UVB-induced apoptosis in normal human keratinocytes [J]. Journal of Biological Chemistry, 2006, 281 (26): 17999-18007.

[21] FRIDOVICH I. Superoxide anion radical ($O_2^{\cdot-}$ radical anion), superoxide dismutases, and related matters [J]. Journal of Biological Chemistry, 1997, 272 (30): 18515-18517.

[22] KOMAROV D A, SLEPNEVA I A, GLUPOV V V, et al. Superoxide and hydrogen peroxide formation during enzymatic oxidation of DOPA by phenoloxidase [J]. Free Radical Res, 2005, 39 (8): 853-858.

[23] LI X R, WANG B, XU J J, et al. In vitro detection of superoxide anions released from cancer cells based on potassium-doped carbon nanotubes-ionic liquid composite gels [J]. Nanoscale, 2011, 3 (12): 5026-5033.

[24] SUSCHEK C V, MAHOTKA C, SCHNORR O, et al. UVB radiation-mediated expression of inducible nitric oxide synthase activity and the augmenting role of co-induced TNF-alpha in human skin endothelial cells [J]. J Invest Dermatol, 2004, 123 (5): 950-957.

[25] SQUADRITO G L, PRYOR W A. Oxidative chemistry of nitric oxide: The roles of superoxide, peroxynitrite, and carbon dioxide [J]. Free Radical Biology and Medicine, 1998, 25 (4/5): 392-403.

[26] BURNEY S, CAULFIELD J L, NILES J C, et al. The chemistry of DNA damage from nitric oxide and peroxynitrite [J]. Mutat Res-Fund Mol M, 1999, 424 (1/2): 37-49.

[27] PERVEEN A, KHAN H Y, HADI S M, et al. Pro-oxidant DNA breakage induced by the interaction of L-DOPA with Cu(II): A putative mechanism of neurotoxicity [J]. Adv Exp Med Biol, 2015, 822: 37-51.

[28] XIAO H, ZHANG W, LI P, et al. Versatile fluorescent probes for imaging the superoxide anion in living cells and in vivo [J]. Angew Chem Int Ed Engl, 2020, 59 (11): 4216-4230.

[29] LI R Q, MAO Z Q, RONG L, et al. A two-photon fluorescent probe for exogenous and endogenous superoxide anion imaging in vitro and in vivo [J]. Biosens Bioelectron, 2017, 87: 73-80.

[30] LEE E R, KIM J H, KANG Y J, et al. The anti-apoptotic and anti-oxidant effect of eriodictyol on UV-induced apoptosis in keratinocytes [J]. Biol Pharm Bull, 2007, 30 (1): 32-37.

[31] TADA M, KOHNO M, NIWANO Y. Scavenging or quenching effect of melanin on superoxide anion and singlet oxygen [J]. J Clin Biochem Nutr, 2010, 46 (3): 224-228.

[32] GRANGE P A, CHEREAU C, RAINGEAUD J, et al. Production of superoxide anions by keratinocytes initiates P. acnes-induced inflammation of the skin [J]. PLoS Pathogens, 2009, 5 (7): e1000527.

[33] CARTER W O, NARAYANAN P K, ROBINSON J P. Intracellular hydrogen-peroxide and superoxide anion detection in endothelial-cells [J]. J Leukocyte Biol, 1994, 55 (2): 253-258.

[34] LEE D K, JANG H D. Carnosic acid attenuates an early increase in ROS levels during adipocyte differentiation by suppressing translation of NOX4 and inducing translation of antioxidant enzymes [J]. Int J Mol Sci, 2021, 22 (11): 6096.

[35] ABUBAKER A A, VARA D, CANOBBIO I, et al. A novel flow cytometry assay using dihydroethidium as redox-sensitive probe reveals NADPH oxidase-dependent generation of superoxide anion in human platelets [J]. Thromb Res, 2018, 164: S229-S230.

[36] BURNAUGH L, SABEUR K, BALL B A. Generation of superoxide anion by equine spermatozoa as detected by dihydroethidium [J]. Theriogenology, 2007, 67 (3): 580-589.

[37] SCHWEIKL H, GALLORINI M, FORSTNER M, et al. Flavin-containing enzymes as a source of reactive oxygen species in HEMA-induced apoptosis [J]. Dent Mater, 2017, 33 (5): E255-E271.

[38] MENG X Q, ZHANG W, ZHANG F, et al. Solanine-induced reactive oxygen species inhibit the growth of human hepatocellular carcinoma HepG2 cells [J]. Oncol Lett, 2016, 11 (3): 2145-2151.

[39] KORSHUNOV S, IMLAY J A. Detection and quantification of superoxide formed within the periplasm of escherichia coli [J]. J Bacteriol, 2006, 188 (17): 6326-6334.

[40] SANDERS S P, HARRISON S J, KUPPUSAMY P, et al. A comparative study of EPR spin trapping and cytochrome c reduction techniques for the measurement of superoxide anions [J]. Free Radic Biol Med, 1994, 16 (6): 753-761.

[41] HERRLING T, ZASTROW L, FUCHS J, et al. Electron spin resonance detection of UVA-induced free radicals [J]. Skin Pharmacol Appl, 2002, 15 (5): 381-383.

[42] SHI Z Z, WU X S, GAO L X, et al. Electrodes/paper sandwich devices for in situ sensing of hydrogen peroxide secretion from cells growing in gels-in-paper 3-dimensional matrix [J]. Anal Methods-Uk, 2014, 6 (12): 4446-4454.

[43] MA X Q, HU W H, GUO C X, et al. DNA-templated biomimetic enzyme sheets on carbon nanotubes to sensitively in situ detect superoxide anions released from cells [J]. Advanced Functional Materials, 2014, 24 (37): 5897-5903.

[44] PRIETO-SIMON B, CORTINA M, CAMPAS M, et al. Electrochemical biosensors as a tool for antioxidant capacity assessment [J]. Sensors and Actuators B-Chemical, 2008, 129 (1): 459-466.

[45] SAKAGAMI H, SATOH K, MAKINO Y, et al. Effect of alpha-tocopherol on cytotoxicity induced by UV irradiation and antioxidants [J]. Anticancer Res, 1997, 17 (3C): 2079-2082.

[46] JURKIEWICZ B A, BISSETT D L, BUETTNER G R. Effect of topically applied tocopherol on ultraviolet radiation-mediated free-radical damage in skin [J]. J Invest Dermatol, 1995, 104 (4): 484-488.

[47] SATOH K, KADOFUKU T, SAKAGAMI H. Effect of trolox, a synthetic analog of alpha-tocopherol, on cytotoxicity induced by UV irradiation and antioxidants [J]. Anticancer Res, 1997, 17 (4A): 2459-2463.

[48] ROSSATO M F, TREVISAN G, WALKER C I B, et al. Eriodictyol: A flavonoid antagonist of the TRPV1 receptor with antioxidant activity [J]. Biochem Pharmacol, 2011, 81 (4): 544-551.

[49] LEE E R, KIM J H, CHOI H Y, et al. Cytoprotective effect of eriodictyol in UV-irradiated keratinocytes via phosphatase-dependent modulation of both the p38 MAPK and Akt signaling pathways [J]. Cell Physiol Biochem, 2011, 27 (5): 513-524.

5 电化学生物传感器在细胞 NO 检测中的应用研究

5.1 引 言

生物体内绝大多数分子是自由基又叫游离基，其化学性质非常活泼，半衰期短，几乎可以在任何惰性条件下和任何惰性物质发生连锁反应，也可与其他物质反应生成新的自由基，从而导致基质的大量消耗及多种自由基产物的生成。人体内的自由基分为氧自由基和非氧自由基。氧自由基占主导地位，大约占自由基总量的95%。在生物体系中主要遇到的是氧自由基，它是人体内最重要的自由基，来自线粒体，通常氧自由基也称为活性氧，包括超氧阴离子自由基、羟自由基、脂氧自由基、二氧化氮、一氧化氮、过氧化氢、单线态氧和臭氧，它们统称为活性氧。体内活性氧自由基具有一定的生理调节功能，如参与免疫调节和信号转导过程[1,2]。但是过多的活性氧自由基就会对机体产生破坏，导致人体正常细胞和组织的损坏，从而引起多种疾病，如心脏病、阿尔茨海默病、帕金森病和肿瘤[3-6]。此外，外界环境中的太阳光辐射、空气污染、吸烟、农药等都会使人体产生更多活性氧自由基，促使核酸突变，这是人类衰老和患病的根源[7]。

一氧化氮（NO）作为典型的 RNS 或 ROS 分子，是一种普遍存在于细胞内的信使分子[8-10]，包含硝酰阴离子（HNO）、亚硝鎓离子（NO^+）、高价态氮氧化物（N_xO_y）、S-亚硝基硫醇（RSNO）、亚硝酸离子（NO^{2-}）和过氧硝酸盐（$ONOO^-$）等[11,12]。NO 是一个具有多重功效的分子，它参与人体内多种重要的生理和病理过程[13-15]。NO 在细胞内硝化压力中扮演着双重角色：一方面细胞产生的 RNS 主要为 NO；另一方面，NO 是其他含氮活性分子的主要来源。NO 通常由 L-精氨酸和氧气在一氧化氮合成酶（NOS）的作用下生成的，然后迅速在细胞内扩散或穿越细胞膜，完成第二信使的任务。例如，它可刺激平滑肌细胞，引起血管舒张[8]，如图 5.1 所示。此外，NO 还可调节很多生理过程包括中枢神经系统、心血管频率、胃肠道活动、泌尿生殖系统、免疫过程、帕金森病及哮喘等[9-12]。然而，在病理情况下，由于 NO 对其他自由基，例如超氧负离子（$O_2^{\cdot-}$）的活性很高，它会转变为对细胞毒性很强的物质。尽管低浓度的 NO 和 $O_2^{\cdot-}$ 在正常生理环境中是无毒的，但是当这两种活性自由基的平衡浓度被打破时，会引起生物体内细胞机能的紊乱。这种失衡主要是由 $O_2^{\cdot-}$ 的产量增加引起消耗量减少，或者 NO 产量减少引起消耗量增加，NO 与 $O_2^{\cdot-}$ 在酶的催化作用下反应生成 $ONOO^-$，它是一种强氧化剂可与大部分生物分子反应导致细胞损伤[13]。NO 在不同的生理过程中表现出两种不同的生物学作用，而且它的水平高低与癌症的发生与发展密切相关[16,17]。目前研究认为低水平的 NO 可以促进细胞的增殖、侵袭和迁移[18,19]，然而高水平的 NO 能够发挥抗肿瘤的作用效果，它主要是通过介导细胞产生毒性，促使氧化/硝基化应激或者 DNA 损伤的发生[20,21]。由此可知，NO 在癌症发生与发展中的作用是明确的。因此，原位实时检测细胞释放的 NO 对于深入探索其在生物体系中的多样作用是很重要的。

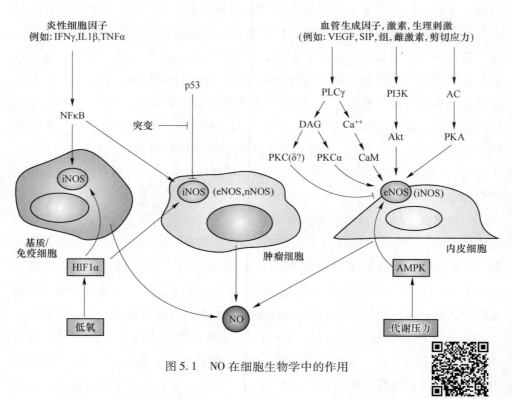

图 5.1　NO 在细胞生物学中的作用

活性氧的化学性质非常活泼，半衰期短，因此其检测方法需要具有高灵敏性和时效性。由于电化学方法不仅具有高灵敏性和低检测限，而且能够对细胞释放的活性氧进行原位实时和连续性检测，已成为检测活性氧的首选方法。在前期实验的基础上，我们通过构纳米复合材料电化学生物传感器实现了对细胞释放的超氧阴离子的检测，这为细胞释放活性氧的检测提供了新的参考依据。因此，本章节主要介绍电化学生物传感器用于检测细胞中 NO 的相关研究。

5.2 rGO/CeO₂纳米复合材料电化学生物传感器的构建

电化学传感器简易、成本低、仪器小型化，更易达到 NO 的实时检测，更适用于 NO 的实时检测同时可以避免对细胞代谢及相关生理过程的破坏和扰乱。高灵敏的检测平台对于实时检测低浓度的 NO 是极其重要的。石墨烯，一种共价键合的单层碳原子二维片状结构材料，具有优良的导电性及易功能化等物理化学性质，已广泛应用于电化学传感器的电极材料。近来，稀土元素氧化物二氧化铈（CeO_2）引起了科学家极大的研究兴趣，CeO_2 拥有独特的 $4f$ 壳层电子结构，可形成多种过渡态，表现出卓越的电化学性能[22,23]。研究表明，CeO_2 对 NO 表现出很强的吸附性能和催化活性[82,24]，有望用于检测 NO 的电极材料。

电化学 NO 传感器三电极体系是由纳米功能材料修饰的工作电极，$Hg/HgCl_2/KCl$ 参比电极和铂丝对电极组成。氧化石墨烯是天然的石墨经过 Hummers 方法进行化学修饰处理得到的。一氧化氮电化学传感器的电极材料是通过还原的石墨烯和二氧化铈纳米复合材料采用水热法制备[25]。构建 NO 电化学传感器的还原石墨烯二氧化铈（rGO/CeO₂）纳米材料是通过水热法进行合成，其主要步骤如图 5.2 所示。首先，取 900 mg 的聚乙烯吡咯烷酮（polyvinylpyrrolidone，PVP），400 mg 的六水合硝酸铈和 7.5 mg 的氧化石墨烯溶解到 30 mL 的二次水中搅拌 30 min。其次，将混合溶液倒入高压反应釜中，在 180 ℃下加热 24 h 后，再经过 70 ℃维持 3 h 进行干燥，获得的 rGO/CeO₂ 纳米复合材料，通过高分辨率电镜对材料的微观结构进行表征，如图 5.3 所示[25]。最后，将合成的 rGO/CeO₂ 纳米复合材料溶解到水中得到 10 mg/mL 的水溶液，室温保存以备用。取 5 μL 的 rGO/CeO₂ 纳米功能材料修饰到预先抛光的玻碳电极上，室温晾干以备用。我们所构建的 NO 电化学生物传感器检测平台如图 5.4 所示，该检测平台是由三电极体系（工作电极、对电极和参比电极）构成，进而实现原位监测细胞释放的 NO 浓度。

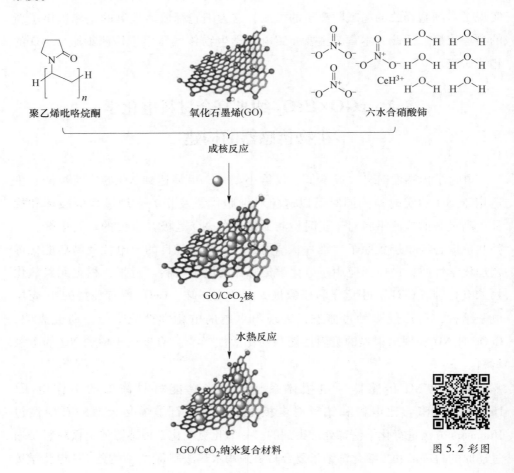

图 5.2　rGO/CeO$_2$ 纳米复合材料的制备过程

电化学检测细胞中 NO 水平实验操作流程如下：首先，将细胞密度大约为 4×10^5/mL 种植于 35 mm 培养皿中，放置于 37 ℃、5% 的二氧化氮细胞培养箱中培养过夜。当细胞密度达到 70%~80% 时选用索拉非尼和一氧化氮抑制剂（L-NMMA）对细胞进行不同处理。其次，依据处理的不同时间段，将培养皿取出放置于预先修饰的 rGO-CeO$_2$ 电极下进行检测。最后，通过软件对数据处理分析。

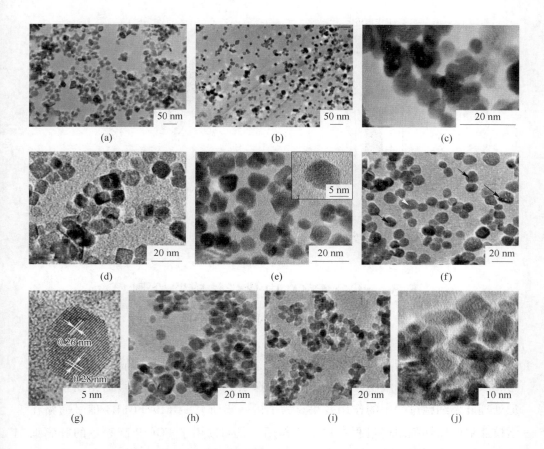

图 5.3　rGO/CeO$_2$ 纳米复合材料的透射电镜图片

(a) CeO$_2$；(b) 掺杂比例为 Ce^{3+}/OH$^-$ = 1∶3 合成 rGO/CeO$_2$ 纳米复合材料的形貌；

(c) 掺杂 Ce^{3+}/OH$^-$ 比例为 2∶1 合成 rGO/CeO$_2$ 的复合材料图片；

(d) 掺杂 Ce^{3+}/OH$^-$ 比例为 1∶1 合成 rGO/CeO$_2$ 的复合材料图片；

(e) 掺杂 Ce^{3+}/OH$^-$ 比例为 1∶2 合成 rGO/CeO$_2$ 的复合材料图片；

(f) 掺杂 Ce^{3+}/OH$^-$ 比例为 1∶3 合成 rGO/CeO$_2$ 的复合材料图片；

(g) Ce^{3+}/OH$^-$ = 1∶3 合成 rGO/CeO$_2$ 纳米复合材料的高分辨率电镜图片；

(h) 掺杂 Ce^{3+}/OH$^-$ 比例为 1∶4 合成 rGO/CeO$_2$ 的复合材料图片；

(i) 掺杂 Ce^{3+}/OH$^-$ 比例为 1∶5 合成 rGO/CeO$_2$ 的复合材料图片；

(j) 掺杂 Ce^{3+}/OH$^-$ 比例为 1∶10 合成 rGO/CeO$_2$ 的复合材料图片

图 5.3 彩图

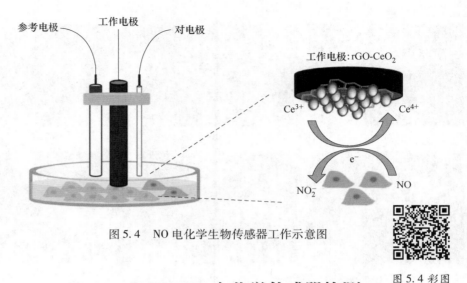

图 5.4 NO 电化学生物传感器工作示意图

图 5.4 彩图

5.3 rGO-CeO$_2$ 电化学传感器检测肝癌细胞中 NO 的水平

本课题通过生物学和电化学方法相结合研究索拉非尼对肝癌细胞的生存能力、氧化应激状况和细胞凋亡的影响,来探索 FGF19/FGFR4 信号通路在索拉非尼处理肝癌细胞过程中的作用效果。为了进一步研究 FGF19/FGFR4 信号通路在索拉非尼肝癌细胞耐药过程中的重要作用,我们选用了 FGF19 高表达的肝癌细胞构建了索拉非尼耐药细胞系,通过探讨不同水平 FGF19 以及在阻断该信号通路的条件下,验证 FGF19/FGFR4 信号通路是否参与肝癌细胞的耐药和在肝癌耐药细胞中的作用机制。上述研究为解决临床中索拉非尼在肝癌病人中的耐药问题提供了一种潜在的治疗措施。

5.3.1 基于 rGO-CeO$_2$ 的 NO 电化学传感器的校正

NO 与血红蛋白结合生成铁亚硝酰血红蛋白或 S-亚硝基血红蛋白,两种反应产物可以掩蔽 NO 的生物活性[26,27]。图 5.5(a)中蓝线表示含有 0.25 nmol/L NO 时,测试系统在 0.85 V 的电位处出现较明显的氧化峰。图 5.5(a)中的红线则表示 1.5 mmol/L 血红蛋白用于消除 NO 的生物活性,可得 NO 特征峰电流值急剧下降。结果表明,该纳米功能材料电极可用于特异性检测 $O_2^{\cdot-}$ 和 NO。为了进一步证明该传感器的检测灵敏度,使用时间电流曲线

对一系列 $O_2^{·-}$ 和 NO 的浓度进行测定。电化学传感器在 5 s 内可以对 $O_2^{·-}$ 做出快速的响应,而且具有 10~200 nmol/L 的稳定检测范围。由超氧阴离子的校准曲线可获得传感器具有 2.5 nmol/L 的检测极限和 3.95 nA/(nmol·L^{-1}) 的敏感度(见图 5.5(b)插图)。从图 5.5(b)浓度校正曲线可以计算出 NO 传感的响应时间为 4 s,灵敏度为 0.069 μA/(μmol·L^{-1}),检测范围为 0.2~4 μmol/L 和 28.17 nmol/L 的检测限。

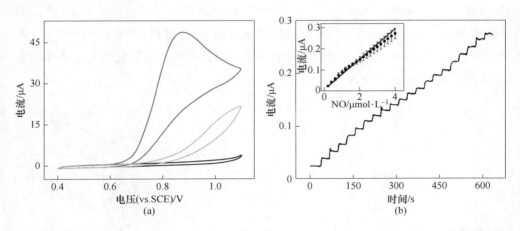

图 5.5 rGO-CeO_2 修饰玻碳电极检测 NO(循环伏安曲线检测 PBS 为黑色线,PBS 加 250 nmol/L NO 为蓝色线,PBS 加 250 nmol/L NO 和 1.5 mmol/L 的血红蛋白为红色线)(a) 和不同浓度 NO 的时间电流曲线和浓度校正曲线(b)

图 5.5 彩图

5.3.2 基于 rGO-CeO_2 的 NO 电化学传感器的稳定性与选择性

采用计时电流法对 1.2 μmol/L NO 进行间歇性检测,研究了该传感器的稳定性。该电极不使用时,悬于 4 ℃ 的 0.01 mol/L PBS 上方保存,20 天后对相同浓度的底物响应可保持为初始时的 90%,说明此电极具有良好的贮存稳定性。此外,将该传感器在 PBS 中连续扫描 20 圈,峰电流基本没有发生变化,说明此电极有良好的操作稳定性。

电化学传感器很容易受到其他化合物的干扰,这会导致测量结果的不准确,例如,在测量某种化合物的浓度时,如果存在其他化合物干扰,可能会引起误差。由于抗坏血酸(AA)、多巴胺(DA)和尿酸(UA)等生物小分子在人体中的含量与多种疾病相关。因此,为了排除其对 rGO-CeO_2 传感器的干扰,研究了

rGO-CeO$_2$ 的选择性和抗干扰能力,对二者进行了检测[25]。实验结果表明,对于 rGO-CeO$_2$ 传感器,6.0×10^{-5} mol/L Ca^{2+}、K$^+$、Na$^+$、CO$_3^{2-}$、NO$_3^-$、Cl$^-$ 和 1.0×10^{-6} mol/L 的尿酸(UA)不会对 6.0×10^{-7} mol/L NO 引起任何干扰(见图 5.6(a))。据报道,在生理环境中,多巴胺 DA 的浓度很低,为 $1.0\times10^{-7}\sim1.0\times10^{-6}$ mol/L[28]。Kutnink 等人的研究结果表明,白细胞中 AA 的平均浓度为 $(2.43\pm1.63)\times10^{-6}$ g/10^8 个细胞(大约为 2.3×10^{-6} mol/L)[29]。通过抗感染实验检测发现 2.3×10^{-6} mol/L AA 和 1×10^{-6} mol/L DA 对 4.3×10^{-6} mol/L 的 NO(细胞密度为 10^5/mL 时,活细胞释放的 NO)没有产生明显的干扰(见图 5.6(b))。综上可知,该传感器表现出了良好选择性和抗干扰能力,并可用于 rGO-CeO$_2$ 后续细胞释放 NO 的检测。

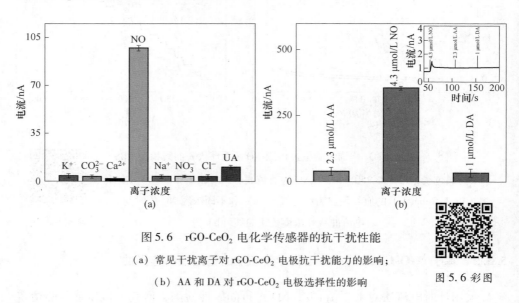

图 5.6 rGO-CeO$_2$ 电化学传感器的抗干扰性能

(a) 常见干扰离子对 rGO-CeO$_2$ 电极抗干扰能力的影响;

(b) AA 和 DA 对 rGO-CeO$_2$ 电极选择性的影响

图 5.6 彩图

5.3.3 索拉非尼在抑制肝癌细胞增殖过程中诱导细胞产生 NO

肝细胞癌是主要来源于肝细胞的肿瘤,它的致死率位居世界第三。我们前期的工作中已经介绍了 FGF19 是一个致癌基因能够负调控酪氨酸激酶,然而索拉非尼的作用效果及其具体的分子机制至今仍然是未知的。索拉非尼作为酪氨酸激酶抑制剂其治疗目的是提高肝癌病人的成活率[30-33]。但是,索拉非尼作为临床一线治疗药物仅能够延长病人 3 个月的生存周期[34],因此,探究索拉非尼关键的分子机制对了解索拉非尼的作用效果和耐药机理具有重要作用。此外,新的分子

5.3 rGO-CeO$_2$ 电化学传感器检测肝癌细胞中 NO 的水平

靶标的设计与索拉非尼的联合用药对于提高肝癌病人的生存质量是一个主要解决途径。因此，本工作通过 rGO-CeO$_2$ 电化学传感器评价 FGF19 和索拉非尼对肝癌细胞生长情况的影响。

FGF19 是一个回肠激素，它是由小肠的细胞分泌而且能够延长一些胰岛素的效率[35,36]。已有报道表明，FGF19 通过激活其相应的受体 FGFR4 在癌症的发生和发展中发挥关键的作用[36,37]。FGF19 基因和 FGFR4 抑制剂的联合应用已经被发现能够控制肝癌细胞中的多种致癌的信号通路，其中 FGF19/FGFR 在肝癌的上皮样细胞中可以诱导 GSK3β/β-catenin/E-cadherin 信号通路而且能够促进上皮间质转化和侵袭[37]。最近新一代基于索拉非尼响应情况的 DNA 拷贝数的序列分析揭示出索拉非尼与 FGF19 的改变具有相关性，该分析认为 FGF19 的扩增或许是索拉非尼响应中的一个重要预测[38]。我们发现 FGF19 在索拉非尼耐药肝癌细胞诱导氧化应激介导的凋亡中发挥重要的作用[39]，同时也发现 FGF19 在索拉非尼的作用过程和索拉非尼肝癌耐药细胞中发挥主要作用。

为了验证肝癌细胞对索拉非尼的敏感性，我们选用了不同浓度的索拉非尼处理 MHCC97L、MHCC97H、SMCC-7721 和 HepG2 细胞 72 h，通过细胞增殖实验检测了细胞在药物处理下的生长能力。我们发现索拉非尼能够有效抑制肝癌细胞的增殖能力，并且优化出细胞的半数致死浓度为 4 μmol/L（见图 5.7(a)），我们将选用这一浓度用于本工作的后续实验研究。事实已经证明 NO 在调节索拉非尼抑制癌细胞的增殖方面发挥重要的调节作用，然而对于二者之间的具体调节关系还不清楚。因此，我们对索拉非尼是否能够通过促进 NO 的产生进行抑制细胞的增殖进行了研究。首先，种植适宜密度的细胞，细胞培养过夜后，选用 4 μmol/L 索拉非尼处理肝癌细胞系，处理时间分别为 0 h，8 h，12 h 和 24 h。为了验证索拉非尼处理后产生的是 NO，我们选用 NO 清除剂 L-NMMA 作为参照，从实验结果可知索拉非尼可以诱导肝癌细胞产生 NO。选用 DAF-FM DA 荧光探针对索拉非尼处理组和对照组细胞产生 NO 的水平进行荧光染色。从实验中我们发现在药物处理细胞 8 h 时能够刺激细胞产生高水平的 NO，药物继续处理达到 24 h 时，细胞释放的 NO 量逐渐降低。在荧光倒置显微镜下观察细胞内 NO 的荧光强度变化，并获取细胞图片进行统计学分析（见图 5.7(b)）。

通过 NO 电化学传感器的检测，我们发现在前 8 h 索拉非尼处理肝癌细胞时可以引起细胞产生 NO 的水平逐渐增高，然后呈现下降趋势，当药物处理 48 h 时细胞释放的 NO 水平已接近于药物处理前水平（见图 5.8）。荧光染色实验结果进

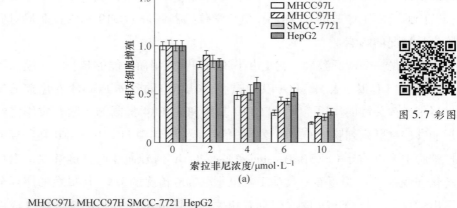

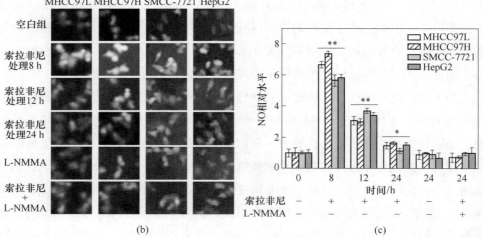

图 5.7 索拉非尼诱导 NO 产生并抑制肝癌细胞的增殖

（a）不同浓度的索拉非尼处理 MHCC97L、MHCC97H、SMCC-7721 和 HepG2 细胞 72 h 后，通过细胞增殖试剂盒检测细胞的增殖能力；（b）MHCC97L、MHCC97H、SMCC-7721 和 HepG2 细胞经过 4 μmol/L 的索拉非尼处理不同时间点后由 DCFH-DA 荧光染料检测细胞 NO 的释放水平并进行统计学分析；（c）(b) 图荧光强度数据的统计图

一步证实了这一发现，索拉非尼能够诱导细胞产生高水平的 NO，荧光染色实验与电化学检测结果具有相似的趋势。通过上述的实验我们认为索拉非尼诱导细胞产生 NO 或许与它抑制增殖的效率具有密切关系。

5.3.4　FGF19 影响索拉非尼刺激肝癌细胞产生的 NO 水平

前期的工作中，我们已经阐述了 FGF19 在 MHCC97L 细胞中表达水平低于

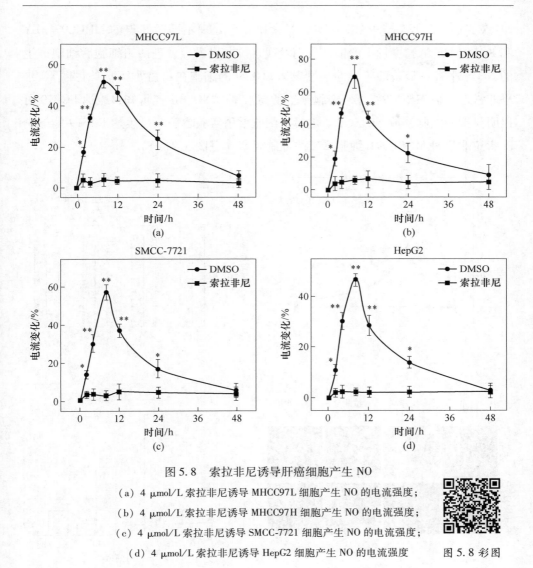

图 5.8　索拉非尼诱导肝癌细胞产生 NO

(a) 4 μmol/L 索拉非尼诱导 MHCC97L 细胞产生 NO 的电流强度；
(b) 4 μmol/L 索拉非尼诱导 MHCC97H 细胞产生 NO 的电流强度；
(c) 4 μmol/L 索拉非尼诱导 SMCC-7721 细胞产生 NO 的电流强度；
(d) 4 μmol/L 索拉非尼诱导 HepG2 细胞产生 NO 的电流强度

图 5.8 彩图

MHCC97H、SMCC-7721 和 HepG2 三种细胞，此研究中选用的敲减细胞系（shNC 和 shFGF19）和过表达细胞（EV，FGF19 O/E）均来自前期研究工作，并已经进行了验证[37]。有趣的是，在此我们通过细胞增殖实验结果还发现与 MHCC97H、SMCC-7721 和 HepG2 三种细胞相比，MHCC97L 细胞对于高剂量的索拉非尼（6 μmol/L 和 10 μmol/L）较为敏感（见图 5.7(a)），这一发现促使我们研究 FGF19 是否参与了索拉非尼诱导细胞产生 NO 的过程。为了探索 FGF19 在索拉非尼处理的肝癌细胞中的作用，通过 NO 电化学传感器（见图 5.9(a)）和细胞内

NO 荧光染色（见图 5.9(b)）实验对索拉非尼处理不同时间点的 MHCC97L EV 和 FGF19 OE 细胞进行了评估，我们发现索拉非尼诱导处理两组细胞后细胞产生的 NO 水平存在显著的差异性，过表达 FGF19 的细胞在药物处理下产生的 NO 水平非常低，而 MHCC97L EV 细胞产生了高水平的 NO，由此可知过表达 FGF19 可以明显抑制细胞释放 NO 水平。随后，细胞增殖实验也表明过表达 FGF19 能够消除索拉非尼对 MHCC97L 细胞增殖的抑制效率（见图 5.9(c)）。

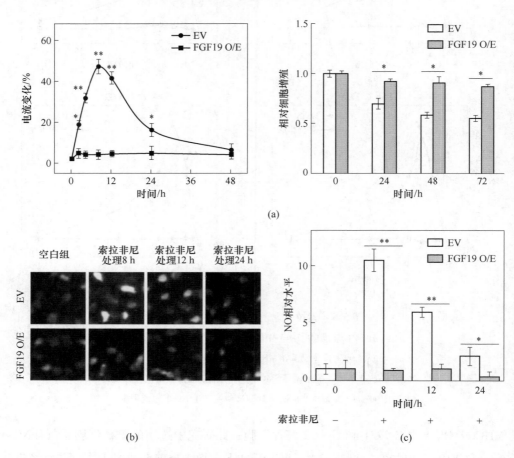

图 5.9　过表达 FGF19 抑制索拉非尼诱导细胞产生 NO 并且提高细胞增殖能力

（a）过表达 FGF19 和对照组细胞在 4 μmol/L 索拉非尼处理不同时间点通过电化学生物传感器检测细胞释放的 NO 水平；(b) 索拉非尼处理不同时间点后细胞内 NO 产生的荧光强度，同时我们选用 NIH Image J 对细胞荧光强度进行了量化；(c) 细胞增殖实验是选用细胞增殖试剂盒进行并获得实验结果

图 5.9 彩图

反之,我们通过慢病毒载体在高表达 FGF19 基因的 MHCC97H 细胞中敲减 FGF19 构建了敲减 FGF19(shFGF19)和空载体细胞系(shNC)。然后,选用 4 μmol/L 索拉非尼对 shFGF19 和 shNC 细胞处理一系列时间点后进行电化学检测两组细胞在处理后细胞释放的 NO 水平是否具有差异性,从而进一步说明 FGF19 表达水平的高低影响细胞抵抗索拉非尼耐受能力。与我们的设想一致的是,敲减 FGF19 后可以增加索拉非尼诱导细胞释放 NO 的水平(见图 5.10(a)),同时细胞内 NO 荧光染色实验结果也表明与对照组细胞相比细胞内的 NO 水平也显著增加(见图 5.10(b))。根据目前的研究可知,NO 水平升高可能对细胞的增殖产生

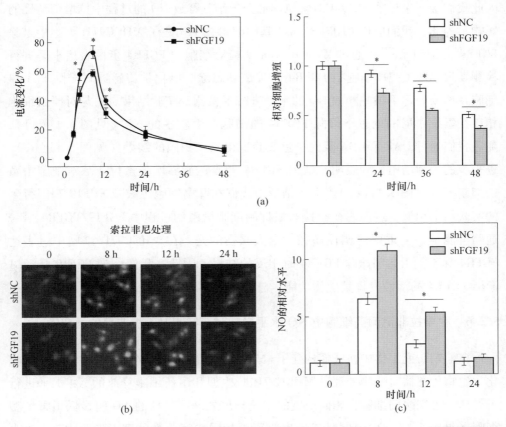

图 5.10 敲减 FGF19 增强肝癌细胞对索拉非尼的敏感性

(a)电化学生物传感器检测细胞释放的 NO 水平;(b)索拉非尼处理不同时间点后细胞内 NO 产生的荧光强度,同时选用 NIH Image J 对细胞荧光强度进行了量化;(c)细胞增殖试剂盒分析 shFGF19 对细胞增殖的影响

图 5.10 彩图

影响进而引起细胞的死亡，在此我们选用了细胞增殖实验对敲减 FGF19 和对照组细胞的增殖能力进行检测，通过对实验结果进行分析可知敲减 FGF19 能够显著降低细胞的增殖能力（见图 5.10(c)）。总之，从这些发现可以揭示出 FGF19 通过下调细胞内 NO 水平达到抑制索拉非尼的抗增殖活性进而影响索拉非尼治疗 HCC 细胞的效果。

5.3.5　NO 电化学传感器检测肝癌细胞释放的 NO 水平

目前的研究已经证实 FGF19 通过激活它的特异性受体 FGFR4 来发挥其功能，因此我们接下来评估 FGFR4 对索拉非尼介导细胞释放 NO 的过程中抗增殖活性的影响。为此，我们应用 CRISPR-Cas9 基因编辑技术构建了 MHCC97H 细胞中敲除 FGFR4 的细胞系[37]，通过 Western Blot 实验对细胞中 FGFR4 蛋白表达水平进行检测发现 MHCC97H 细胞中 FGFR4 已被成功敲除。电化学传感器对敲除 FGFR4 细胞在药物处理下细胞释放 NO 的水平进行了检测，实验结果表明与对照组细胞相比在索拉非尼的处理下敲除 FGFR4 的细胞产生较多的 NO（见图 5.11(a)），随后我们通过 DAF-FM DA 荧光染色实验也得到了相同的结果（见图 5.11(b)）。进一步的细胞增殖实验表明，缺失 FGFR4 的表达能够增强索拉非尼对细胞增殖的抑制率（见图 5.11(c)）。上述结论与我们在肝癌细胞中敲减 FGF19 基因得到的实验结果相似，这说明在索拉非尼抑制细胞增殖的过程中 FGF19/FGFR4 信号通路可能是一个重要的调控途径。令人兴奋的是，在 MHCC97H 细胞中过表达 FGFR4 并不能引起与敲除 FGFR4 时相反的结果。因此，我们认为 FGF19 是通过激活 FGFR4 的活性发挥其生物功能，与 FGFR4 的表达水平高低没有关系。

5.3.6　在索拉非尼耐药细胞中 NO 水平影响耐药细胞对索拉非尼的敏感性

在前期研究工作中，我们构建了索拉非尼耐药细胞系 MHCC97H-Sora[39]，电化学生物传感器对药物处理后 MHCC97H 野生型和耐药细胞释放的 NO 水平进行了评估，可知耐药细胞在相同浓度的索拉非尼处理下释放的 NO 明显低于野生型细胞（见图 5.12(a)）。同时我们也监测了相同浓度药物处理下细胞内 NO 的水平，此实验结果与电化学检测结果一致（见图 5.12(b)）。MHCC97H 野生型和耐药细胞在相同的处理条件下，耐药细胞抵抗药物的增殖能力明显高于野生型细胞（见图 5.12(c)），这说明 MHCC97H 野生型细胞与耐药细胞相比，在用高剂量索拉非尼（20 μmol/L）处理下具有较强的增殖能力而且产生较少的 NO，从这

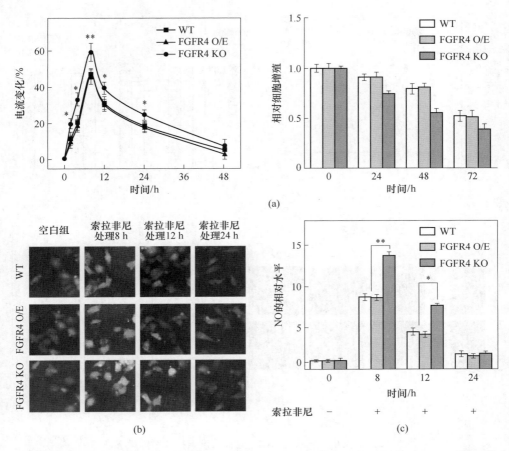

图 5.11 激活 FGFR4 能够调节索拉非尼诱导肝癌细胞
释放 NO 和抗增殖效率

(a) 细胞释放 NO 水平通过电化学生物传感器进行评价；
(b) 细胞释放 NO 水平通过细胞荧光染色实验检测进行评价；
(c) 细胞增殖实验检测细胞的增殖能力

图 5.11 彩图

一实验中我们认为 NO 水平的增高或许能够为克服索拉非尼耐药问题提供新的治疗方法。

5.3.7　电化学方法评价 BLU9931 增强索拉非尼对耐药细胞的抑制能力

由于 BLU9931 是一个高度选择性的 FGFR4 抑制剂，我们决定研究 BLU9931 在索拉非尼耐药细胞中的作用效果。通过不同浓度的 BLU9931 处理细胞后，收

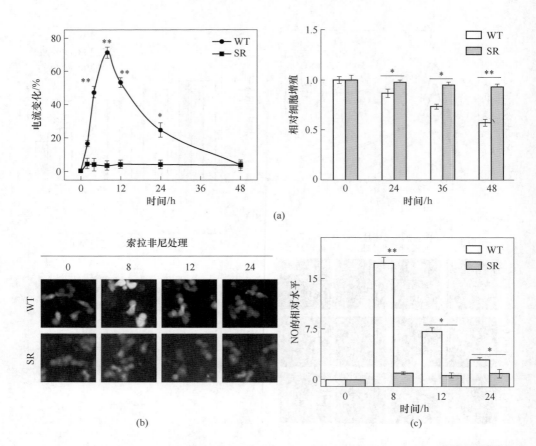

图 5.12 索拉非尼肝癌耐药细胞对高浓度索拉非尼的敏感性，
索拉非尼肝癌耐药细胞培养在含有 1 μmol/L 的索拉非尼
培养基中然后选用 20 μmol/L 的索拉非尼处理
用于验证细胞的敏感性

图 5.12 彩图

(a) 细胞释放 NO 的量用电化学生物传感器进行监测；(b) 细胞内产生 NO 的情况由 DCFH-DA
进行观察并对数据量化；(c) 细胞增殖实验试剂盒检测细胞的增殖能力

集蛋白样品进行 Western Blot 发现当降低 FGFR4 活性时，BLU9931 具有剂量依赖性（见图 5.13(a)），而且它也可以抑制索拉非尼耐药细胞的增殖能力（见图 5.13(b)）。更重要的是，我们发现 BLU9931 和 20 μmol/L 的索拉非尼联合处理索拉非尼耐药细胞能够显著的降低细胞的增殖能力，并且发现 BLU9931 和高浓度的索拉非尼联合使用具有协同作用，二者联合使用效果强于任何一个药物的单独作用效果（见图 5.13(b)）。总而言之，我们发现 BLU9931 处理索拉非尼耐药细胞的效果与 NO 水平的增加具有相关性（见图 5.13(c) 和 (d)）。因此，我们认为高剂量索拉非尼与 BLU9931 联合使用能够增强其抗增殖能力。

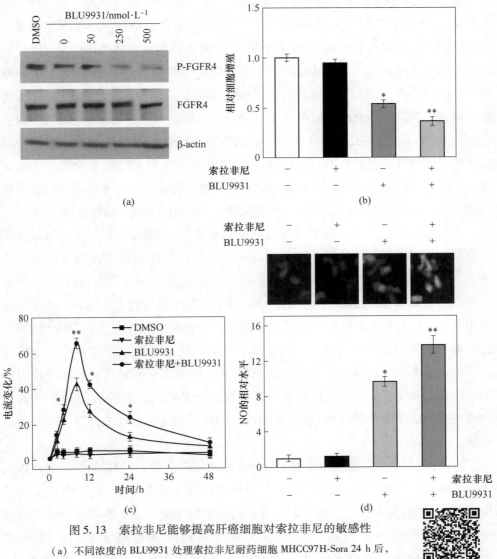

图 5.13 索拉非尼能够提高肝癌细胞对索拉非尼的敏感性

(a) 不同浓度的 BLU9931 处理索拉非尼耐药细胞 MHCC97H-Sora 24 h 后，由 Western Blot 检测磷酸化 FGFR4 蛋白表达水平的变化情况；(b) MHCC97H-Sora 细胞在存在或者缺失 20 μmol/L 索拉非尼处理下同时用 500 nmol/L 的 BLU9931 处理 48 h 后，通过细胞增殖实验试剂盒进行分析细胞的增殖能力；(c) 在此处理条件下细胞产生的 NO 水平由电化学生物传感器进行监测；(d) DCFH-DA 荧光染色实验检测细胞内 NO 的水平

5.3.8 rGO-CeO$_2$ 电化学生物传感器在肝癌治疗中的潜在应用价值

NO 是众多癌细胞病变过程中释放的一种信号传导分子。然而，由于 NO 极易扩散、浓度低、寿命短，实现 NO 的实时检测仍是极大的挑战。本工作通过水热法合成了形貌可控的还原氧化石墨烯-二氧化铈纳米复合物（rGO/CeO$_2$），并用它来构建高灵敏的 NO 实时检测传感器。rGO/CeO$_2$ 纳米复合物中 CeO$_2$ 的晶体形貌对 rGO-CeO$_2$ 传感器的性能有重要影响。其中，六边形的 CeO$_2$ 纳米晶体催化性能最佳，达到了最高的灵敏度（1676.06 mA/(mol·L^{-1}·cm^2)），较宽的线性范围（18.0 nmol/L~5.6 μmol/L）及低的检测限（9.6 nmol/L）。这一最优性能主要归功于特殊形貌的 CeO$_2$ 良好的催化性能及 rGO 优良的导电性和高的比表面积。本研究工作通过有效调控复合物中单组分的优势来合成具有超强协同作用的传感平台，同时也为快速实时检测活细胞释放 NO 提供了极大的可能性。

尽管索拉非尼已经表现出可以提高肝癌后期病人的成活率[40,41]，但是从统计数据可知，由于索拉非尼的耐药问题一直使其在临床应用中面临着巨大的挑战[42-44]。索拉非尼作为酪氨酸激酶抑制剂能够抑制多种信号通路，而且可以激活一些通路开关和补偿通路[30-33]，这为开发新的治疗策略开辟了新途径，例如联合使用索拉非尼与其他的抗癌药物增强索拉非尼的抗肿瘤效率并克服其耐药问题。已有报道表明 FGF19 的表达水平与索拉非尼耐药呈现正相关[38]。因此，我们进一步评估了 FGF19 在索拉非尼处理肝癌细胞的作用效果和耐药性方面的作用机制。研究已表明索拉非尼可以抵消肝癌细胞的增殖能力，我们认为 FGF19 是通过索拉非尼诱导细胞产生 NO 进一步调控它的抗增殖能力，这或许是一个重要的关联。

NO 在癌症中的双重作用主要依赖于它的浓度和肿瘤的微环境[17]。生物体基因的多样性可以编码 NO 合成酶，它与多种癌症的发生密切相关，同时也证明了 NO 与癌症的相关性[45,46]。索拉非尼能够增强肝癌细胞中 NO 的水平，而且可以增强它的抗增殖效率。与这一概念相一致的是，与野生型肝癌细胞相比肝癌耐药细胞中产生较低水平的 NO。NO 可能来自三种不同的亚型，神经元（nNOS/NOS1），诱导型（iNOS/NOS$_2$）和内皮型一氧化氮合成酶[16,45]，进一步的实验需要区别索拉非尼调节的是哪一种 NO 亚型。

相关报道表明，过表达 FGF19 水平可以导致 FGF19/FGFR4 信号通路发生异常，因此，深入理解各种肿瘤从增殖到代谢的发展过程是必须的[37,47]。我们在前期的研究工作中已经阐述了 FGF19 在肝癌细胞中的分泌通过激活 FGFR4 来实现，它通常分为自分泌和旁分泌两种类型，因此，这是一个经典的致癌循环[37]。这个研究工作中，我们进一步理解的另一个机制是索拉非尼处理肝癌细胞后 FGF19/FGR4 信号通路或许可以抑制 NO 的产生，这一发现可以把 FGF19 作为潜在的治疗靶标，来探索索拉非尼与其他药物联合使用在肝癌治疗过程中的分子机制。

FGF19 作为药物靶标已得到一定程度的发展，它通常包括多克隆抗体和小分子药物。目前已发现的 1A6 是一个中和抗体，它能够阻断 FGF19 的媒介直接作用于 FGF19 达到抑制肿瘤的发生[48,49]。由于 FGF19 作为一个多功能的激素应用于调节胆汁酸，碳水化合物和肝脏的再生[50]。无致癌性的变体（M70）通过阻止体内与 FGF19 相关的正常胆汁酸平衡克服 1A6 不理想的治疗效果[49,51]。体内研究已经表明 M70 通过 FGF19 的介导能够抑制肿瘤的增长效率，而不会干涉它在体内胆汁酸平衡中的作用[51]。FGF19 的作用受到干扰也可以使其特异性受体 FGFR4 的作用受到影响。我们应用 FGFR4 的抑制剂 BLU9931 处理索拉非尼耐药细胞系发现它可以有效抑制细胞的增殖并且增强肝癌细胞对索拉非尼的敏感性。

我们的实验结果已表明 FGF19/FGFR4 信号通路的超活化是索拉非尼肝癌耐药细胞的主要作用机制，它的失活通过抑制 NO 相关的增殖能够提高肝癌细胞对索拉非尼的敏感性。接下来我们将要进行的研究是发展一种能够鉴定出那些应用索拉非尼有益的病人，而且需要临床试验中评估索拉非尼与 FGF19 的特异性靶标联合应用于索拉非尼肝癌耐药病人的治疗效果。

为了探索 FGF19 在索拉非尼诱导肝癌细胞产生 NO 中的作用。我们在前期研究的基础上，选择 MHCC97L、MHCC97H、SMCC-7721 和 HepG2 四种肝癌细胞对索拉非尼的敏感性进行了检测，并通过电化学生物传感器和荧光染色实验监测了细胞在索拉非尼处理下释放的 NO 水平。基于前期的实验，我们认为 FGF19 在 NO 的产生中可能发挥调节作用，进而影响索拉非尼作用于病人的治疗效果。当 FGF19 过表达时可以抑制索拉非尼诱导细胞产生 NO 的水平，同时也伴随着抑制肝癌细胞的增殖。然而，当 FGF19 的表达降低或者敲除 FGFR4 的表达时则出现了相反的实验结果。我们也发现通过 BLU9931 使 FGFR4 基因的表达失活能够为克服肝癌细胞索拉非尼的耐药问题提供一个新的策略。综上所述，我们认为 FGF19 或许可以充当一个具有吸引力的治疗靶点，而且阻断 FGF19/FGFR4 信号通路也可以增强索拉非尼在肝癌细胞中的作用效果。

5.4 金纳米颗粒-3D 石墨烯水凝胶纳米复合材料电化学传感器检测细胞释放 NO

一氧化氮（NO）在生物系统中起着多种重要作用，不仅具有神经传递、免疫防御功能和血管松弛等方面的生理特性，还具有促进肿瘤的发生发展进程[52-54]。NO 的原位检测对于深入研究 NO 在生理与病理进展中的作用具有重要参考价值。然而，由于 NO 是一种非常具有化学活性的物质，可以被氧或超氧化物 NO_2^- 或 NO_3^- 离子迅速氧化。它在生理溶液中的半衰期很短，只有 6 s[55,56]。由于活细胞中 NO 的浓度非常低，且细胞中各种代谢小分子物质均会对检测产生干扰物。因此，为了检测活细胞中释放的 NO，需要构建一种灵敏度高、选择性好、抗干扰能力强与响应时间短的电化学传感器。

为了原位检测活细胞中的 NO，微型化的检测平台至关重要，它可以渗透到单个细胞中。各种微电极，如碳纤维[57]或铂[58]或金微电极[59]，已被开发用于检测 NO。然而，微电极仍然存在干扰信号强、信噪比（S/N）低、可重用性差等问题[60,61]。近年来，微型化的微电极并未使其检测灵敏度进一步提高，随着纳米科学的发展，纳米结构优越、比表面积大、导电性好、生物相容性好、催化活性高、化学修饰独特的电化学传感器已被用于实时、原位检测 NO[60,62]，但是高灵敏度、高选择性的电化学传感器仍有很高的需求。石墨烯是由单层碳原子组成的六边形晶格，三维石墨烯水凝胶（3DGH）由于其柔韧、多孔和三维网络化的结构引起了人们的广泛关注，且具有优异的导电性、高比表面积和快速的电荷流动性，有望进一步改善石墨烯的物理和电化学性能。3DGH 除了用于各种重要的电化学应用。此外，其中心空心核和外壁可以有效地吸附和储存气体。然而，由于其碳的性质，3DGH 具有较差的内在电催化活性。由于金纳米颗粒（Au NPs）已被认为是一种有效的氧化 NO 的电催化剂，因此，本研究首次对 Au NPs 进行了研究，并通过一步还原方法构建成金纳米颗粒-3D 石墨烯水凝胶纳米复合材料（Au NPs-3DGH）用于制备 NO 传感器，其具有高灵敏度和选择性，实现了原位检测活体肿瘤细胞释放 NO。研究发现 Au NPs-3DGH 传感器可以检测出正常细胞与肿瘤细胞释放 NO 的显著差异，将为深入探索黑色素瘤发展进程提供理论参考依据。

5.4.1 Au NPs-3DGH 复合材料的制备及其电极修饰

将均匀氧化石墨烯（GO）的水分散体（2.5 mg/mL）密封在聚四氟乙烯内衬的高压釜中，并在 180 ℃加热 12 h[63]。然后在保持产品高柔性结构的前提下，对产品进行冷冻干燥，得到三维模型。石墨烯（G）的合成方法是：在 pH 值为 10.5 mg/mL、0.1 mg/mL 的氧化石墨烯溶液中加入 2 mmol/L 的维生素 C，加热至 90 ℃，搅拌至氧化石墨烯溶液变为棕色或黑色，然后将上述产物离心干燥。通过简单的一步原位还原构建在 3DGH 上的 Au^{3+} 制备 Au NPs-3DGH，其中 3DGH 作为优良的传导基质，而 Au NPs 直接生长在石墨烯水凝胶的三维表面。将冻干后的 3DGH 加入 1 mmol/L 氯金酸溶液中搅拌 15 min，加入 0.01 mol/L 冰冷 $NaBH_4$ 搅拌至淡黄色变为酒红色，过滤得到 Au NPs-3DGH 复合物（见图 5.14）。将 Au NPs-3DGH 溶解于 Nafion 溶液（0.05%）中，制备 5 mg/mL 溶液，将制备好的溶液（3.5 μL）滴下，表面固定在玻碳电极（GCE）表面干燥。

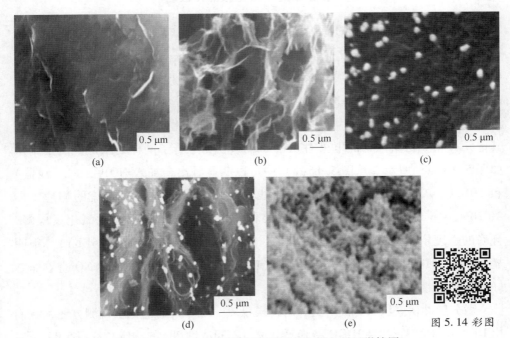

图 5.14 Au NPs-3DGH 复合材料的扫描电子显微镜图

(a) 二维石墨烯片（G）；(b) 三维石墨烯水凝胶（3DGH）；(c) 金纳米粒子/二维石墨烯复合物（Au NPs-G）；(d) 金纳米粒子/三维石墨烯水凝胶（Au NPs-3DGH）；(e) 单独的金纳米粒子（Au NPs）

通过对比二维石墨烯片和三维石墨烯水凝胶（3DGH）发现 3DGH 具有疏松多孔的三维立体状结构，并且具有良好的柔韧性和蓬松度。进一步通过对比金纳米粒子/二维石墨烯复合物（Au NPs-G），金纳米粒子/三维石墨烯水凝胶（Au NPs-3DGH）和单独的金纳米粒子（Au NPs）可知，Au NPs-G 中的金纳米粒子只能分布在石墨烯片的外表面并形成较大的颗粒，Au NPs 由于没有任何基底供其生长，并且团聚到一起形成了很大的聚集体，然而 Au NPs-3DGH 复合纳米材料却非常的疏松多孔，金纳米粒子很均匀地分布在石墨烯水凝胶三维网状结构的内表面和外表面形成了较小的纳米颗粒，并且尺寸均匀分布在 5~40 nm 的范围内。由于 3DGH 具有非常疏松多孔和柔韧蓬松的网状结构以及高浓度的表面活化位点，使得 Au NPs-3DGH 纳米材料形成较好微观结构形态，可以很强地吸附含有 Au(Ⅲ) 的配合物，并在硼氢化钠的还原作用下使 3DGH 上原位生长出较小的纳米颗粒。

5.4.2 Au NPs-3DGH 纳米复合物的一氧化氮催化性能

为了研究材料对 NO 的催化性能，我们使用标准的循环伏安法（CV）来进行分析测定。从图 5.15 中可以明显地看出，Au NPs-3DGH 纳米复合物修饰的电极在当 PBS 溶液中加入了 0.5 mmol/L 的 NO 后，出现了一个很明显且尖锐的氧化峰，证明了 AuNPs-3DGH 纳米复合物对 NO 有着非常好的催化性能。为了更好地分析 Au NPs-3DGH 纳米复合物对 NO 的催化活性，在相同的条件下用其他几种相关材料修饰的电极对 0.5 mmol/L NO 进行了同样的测试，测定结果如图 5.16(a) 所示。从图 5.16(a) 中可以明显看出，加入 NO 后，G（浅蓝）和 3DGH（灰）呈现出平缓的电流曲线而几乎没有氧化反应电流的响应，而 Au NPs、Au NPs-G 和 Au NPs-3DGH 修饰的电极都显示出很大的氧化反应峰，且峰的强度依次为 Au NPs < Au NPs-G < Au NPs-3DGH（见图 5.16(b)）。很明显 Au NPs-3DGH 纳米复合物有着最强氧化峰，进一步证明了它对 NO 有着非常好的催化活性。

微分脉冲伏安法（DPV）是一种高灵敏度、低噪声的电化学检测方法，我们用 DPV 来研究 Au NPs-3DGH 纳米复合物对不同浓度 NO 的响应。从图 5.17 可以明显看出 NO 浓度从 0.4 mmol/L 一直降低到 0.05 mmol/L，Au NPs-3DGH 纳米复合物修饰的电仍能呈现出很好的电流响应，并且峰电流与 NO 的浓度有着很好的线性关系，预示着其可以进行较好的定量分析。

5.4 金纳米颗粒-3D石墨烯水凝胶纳米复合材料电化学传感器检测细胞释放NO

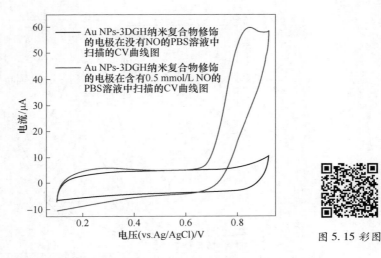

图5.15 彩图

图5.15 不同材料修饰的工作电极的循环伏安曲线（CV）扫描图（一）

（检测条件为：在0.01 mol/L、pH=7.4的磷酸缓冲溶液（PBS）中，扫描速率为50 mV/s）

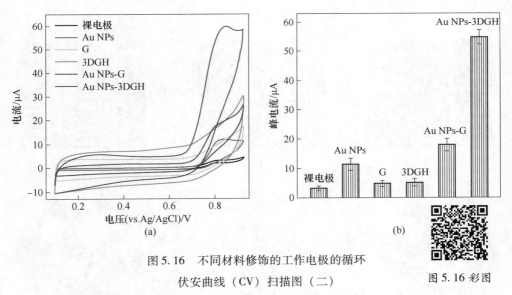

图5.16 彩图

图5.16 不同材料修饰的工作电极的循环伏安曲线（CV）扫描图（二）

(a) 不同对比材料裸电极（黑色）、Au NPs（紫红）、G（浅蓝）、3DGH（灰）、Au NPs-G（深蓝）、Au NPs-3DGH（大红）修饰的电极在0.5 mmol/L NO的PBS溶液中扫描的CV曲线图；
(b) 对应于 (a) 图的五种不同材料修饰的电极对NO的响应

Au NPs-3DGH 纳米复合物对NO的催化机理如图5.18所示。3DGH由于它的碳材料天性而对NO几乎没有催化作用，但是它提供了巨大的比表面积，使得金

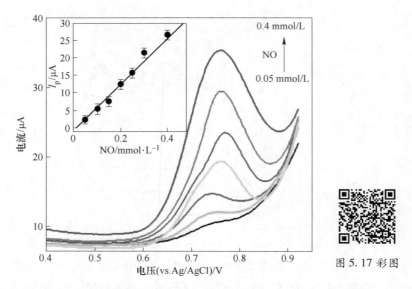

图 5.17 Au NPs-3DGH 修饰的电极对不同浓度 NO 的微分脉冲伏安法（DPV）扫描图
（插图为峰电流与 NO 浓度的关系曲线）

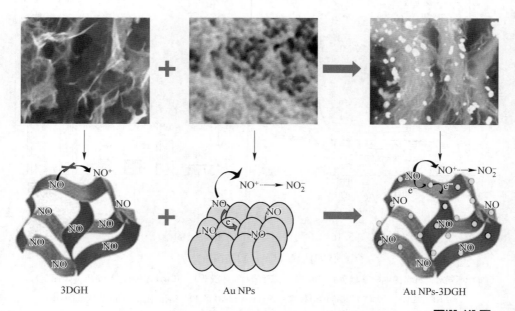

图 5.18 Au NPs-3DGH 纳米复合物对 NO 的催化机理图

纳米粒子能够实现更多量、更小尺寸及更均匀地附着，从而得到更多的活性位点和更快的电子传输。因此，这种独特的 Au NPs-

3DGH 纳米复合物可以使更多的 NO 吸附到金纳米粒子的活化中心，从而快速地失去电子形成 NO^+。由于 Au NPs-3DGH 高度疏松且多孔的结构可以使大量的 NO^- 与 OH 发生反应生成 NO_2，以快速地完成 NO 在电极上的整个反应过程。很显然，Au NPs 和 3DGH 的协同效应促使纳米材料具有非常好的催化性能。

5.4.3 Au NPs-3DGH 纳米复合物传感膜的选择性

我们用计时电流法（i-t）对 Au NPs-3DGH 纳米复合物修饰的电极针对 NO 检测的选择性进行了研究，研究的干扰物是普遍存在于生物体内的其他小分子，例如：NO_2、抗坏血酸（Aa）、多巴胺（Da）等。从图 5.19 中可以明显看出，当分别依次加入相同浓度（1 μmol/L）NO 和其他干扰物质后，Au NPs-3DGH 纳米复合物修饰的电极对 NO 显示出急剧增大的响应电流，而对其他干扰物的响应显得非常微小。表明 Au NPs-3DGH 纳米复合物修饰的电极对 NO 有着很好的选择性。

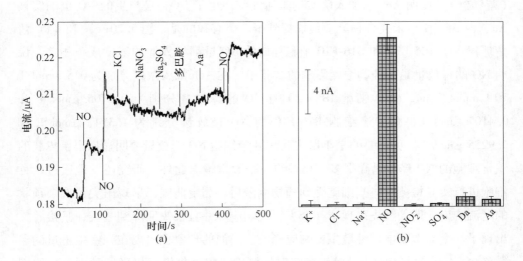

图 5.19　检测 Au NPs-3DGH 电极对 NO 的选择性
（a）Au NPs-3DGH 修饰电极的选择性计时电流曲线；（b）对应于（a）的选择性柱状图

5.4.4 Au NPs-3DGH 纳米复合物传感膜对活细胞的实时监测

我们用 Au NPs-3DGH 纳米复合物修饰的电极作为一个高灵敏的传感器，采用计时电流法分别对培养皿中的小鼠正常皮肤细胞 JB6-C30 和肿瘤细胞 B16-F10

释放的 NO 进行了原位实时的监测，测定条件为保持恒定电压值在 0.81 V（以 Ag/AgCl 为参比电极）。乙酰胆碱（Ach）是目前被广为报道的一种可刺激活细胞产生 NO 的药物，它的刺激原理是能够活化一氧化氮合成酶（NOS），从而产生 NO，而血红蛋白（Hb）是一种典型的 NO 清除剂。为了研究 NO 在黑色素瘤细胞中的重要作用，我们用 Au NPs-3DGH 纳米复合物修饰的电极对小鼠细胞在加入不同浓度 Ach 时，释放 NO 的情况进行了研究，测定结果如图 5.20 所示。从图 5.20(b) 和（e）中可以明显看出，当 1 mmol/L 的 Ach 加入没有细胞的培养基中时，电流几乎没有任何变化（黑），说明 Ach 本身是没有电化学活性的，不会对实验产生干扰。然而当 1 mmol/L Ach 和 1 mmol/L Hb 同时加入分别培养有密度为 1×10^4 的 JB6-C30 细胞（见图 5.20(b)）和 B16-F10 细胞（见图 5.20(e)）中时，产生了一个非常小的电流响应，且很快就降回到基线（浅蓝），这说明 Hb 能够很好地清除由于药物刺激产生的 NO。相反，当 Ach 单独加入培养基中时，JB6-C30 细胞和 B16-F10 细胞的监测曲线都产生了电流的急剧增加（蓝和红），说明 Ach 的加入确实刺激活细胞产生了 NO，且产生的 NO 也确实被 Au NPs-3DGH 纳米复合物灵敏地捕捉到了。更有趣的是，图 5.20(c) 和（f）清楚地揭示了 JB6-C30 和 B16-F10 细胞释放的 NO 量都与 Ach 的浓度成正相关，说明 NO 的释放量是受药物浓度影响的。其中，在 Ach 加入浓度分别为 0.5 mmol/L 和 1 mmol/L 时，正常细胞 JB6-C30 的 NO 的释放量分别为 0.2705 μmol/L 和 0.4107 μmol/L；而肿瘤细胞 B16-F10 的 NO 释放量分别为 1.3147 μmol/L 和 2.9255 μmol/L。更重要的是小鼠皮肤的肿瘤细胞 NO 释放量要明显比小鼠皮肤的正常细胞的 NO 释放量高很多，相差约一个数量级。此外，研究还发现通过两种细胞进行实时检测时，当细胞受到药物刺激后，测定曲线都呈现出尖锐或圆顶状的电流响应，即：首先出现了一个尖锐的响应峰并且很快回降到基线，紧接着又出现了一个类似于电容性质主要响应峰波。类似的现象也出现在一些报道过的谷氨酸盐[64]和一氧化氮[65]的脑细胞实时检测中。根据报道，出现这种现象的原因很可能是由于细胞在受到药物刺激后不仅产生了 NO，也产生了一些具有电活性的副产物。尽管传感器电极具有很好的选择性，由于这些突然产生的活性副产物具有很高的浓度，因此会产生强电流响应。但是尽管如此，它们并没有影响到 NO 的信号响应。因此，测定曲线中才会出现尖锐/圆顶状的电流响应。

为了研究 NO 在病理过程中的重要作用，我们还用 Au NPs-3DGH 纳米复合物修饰的电极对小鼠正常细胞 JB6-C30（蓝）和肿瘤细胞 B16-F10（红）在相同浓

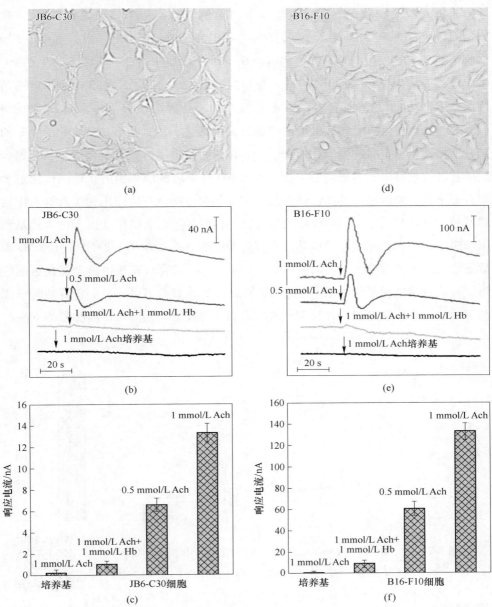

图 5.20　Au NPs-3DGH 纳米复合物传感膜对活细胞产生的 NO 进行实时监测
(a) 小鼠正常皮肤细胞 JB6-C30 在培养皿中的显微镜照片 (细胞浓度为 1×10^5);
(b) Au NPs-3DGH 纳米复合物修饰的电极对 JB6-C30 的实时监测图 (药物在箭头标注的时候加到培养基里); (c) JB6-C30 的响应峰电流的柱状图; (d) 肿瘤细胞 B16-F10 在培养皿中的显微镜照片 (细胞浓度为 1×10^5); (e) Au NPs-3DGH 纳米复合物修饰的电极对 B16-F10 的实时监测图; (f) B16-F10 的响应峰电流的柱状图

图 5.20 彩图

度的药物刺激下释放的 NO 进行原位实时监测,测定结果如图 5.21 所示。当 1 mmol/L 的 Ach 分别加入细胞密度均为 1×10^4 的 JB6-C30 细胞(蓝)和 B16-F10 细胞培养基中时,产生了不同的响应电流。肿瘤细胞 B16-F10(红)产生的 NO 响应电流大约为正常细胞 JB6-C30(蓝)的 5 倍之多。两种细胞在同等浓度的药物刺激下产生的响应电流会有如此大的差别,是因为诱导型一氧化氮合成酶(INOS)在肿瘤细胞中有着比正常细胞更高的表达,因此更多的 INOS 在 Ach 的刺激下便产生了更多的一氧化氮分子。其实很容易理解为什么诱导性一氧化氮合成酶在肿瘤细胞中有着更高的表达,因为 NO 在促进肿瘤细胞的扩增和代谢过程中有着很重要的作用,其中包括直接作用和间接作用,直接作用为引发肿瘤细胞的增殖,迁移和入侵,间接作用为通过影响血管和淋巴管的生成因子的表达从而影响癌细胞的扩增和代谢。因此,在相同浓度的 Ach 刺激下,肿瘤细胞的 NO 释放量要远高于正常细胞。实时监测结果表明了 NO 的释放量很可能与黑色素瘤的形成有着直接地联系。因此,对于 NO 的原位实时监测不仅让我们更好地理解黑色素瘤的重要病理过程,并且很有希望成为一种简单的预知和诊断肿瘤的方法。

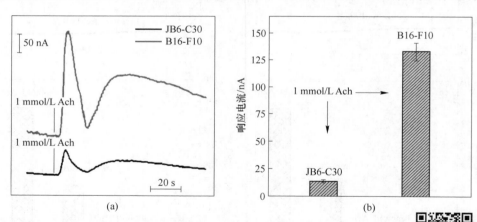

图 5.21 Au NPs-3DGH 纳米复合物修饰的电极对小鼠的 JB6-C30(蓝)
和 B16-F10 细胞(红)在 1 mmol/L Ach 的刺激下的实时监测图
(其中细胞密度为 1×10^4,药物在箭头标注的时间加到培养基里(a)
和对应于(a)的电流响应柱状图(b))

图 5.21 彩图

5.4.5 Au NPs-3DGH 电化学传感器应用展望

综上可知,我们成功开发出了一种简单的方法将金纳米粒子原位地生长到高度疏松多孔的三维石墨烯水凝胶上,形成 Au NPs-3DGH 纳米复合物。它有着很大的电化学活性表面积和很强的一氧化氮催化性能，展现出检测线低、响应时间短以及选择性高的 NO 检测能力。Au NPs-3DGH 纳米复合物同时还显示出非常好的原位实时地监测活细胞释放的 NO 的能力，揭示了肿瘤细胞 B16-F10 产生的 NO 大约为正常细胞 JB6-C30 的 5 倍。对于活细胞释放的 NO 的实时原位检测不仅为更好地理解黑色素瘤提供了重要信息，还有可能用于皮肤肿瘤的预知和诊断。

参 考 文 献

[1] NATHAN C, CUNNINGHAM-BUSSEL A. Beyond oxidative stress: An immunologist's guide to reactive oxygen species [J]. Nat Rev Immunol, 2013, 13 (5): 349-361.

[2] KUSMARTSEV S, NEFEDOVA Y, YODER D. Gabrilovich di: Antigen-specific inhibition of CD8 (+) T cell response by immature myeloid cells in cancer is mediated by reactive oxygen species [J]. J Immunol, 2004, 172 (2): 989-999.

[3] HALLIWELL B. Reactive oxygen species in living systems: Source, biochemistry, and role in human disease [J]. Am J Med, 1991, 91 (3C): 14S-22S.

[4] BENZI G, MORETTI A. Are reactive oxygen species involved in Alzheimer's disease? [J]. Neurobiol Aging, 1995, 16 (4): 661-674.

[5] DRECHSEL D A, PATEL M. Role of reactive oxygen species in the neurotoxicity of environmental agents implicated in Parkinson's disease [J]. Free Radic Biol Med, 2008, 44 (11): 1873-1886.

[6] LIOU G Y, STORZ P. Reactive oxygen species in cancer [J]. Free Radic Res, 2010, 44 (5): 479-496.

[7] SCANDALIOS J G. Oxidative stress: Molecular perception and transduction of signals triggering antioxidant gene defenses [J]. Braz J Med Biol Res, 2005, 38 (7): 995-1014.

[8] LIM M H, WONG B A, PITCOCK W H. et al. Direct nitric oxide detection in aqueous solution by copper (Ⅱ) fluorescein complexes [J]. Journal of the American Chemical Society, 2006, 128 (44): 14364-14373.

[9] BREDT D S, SNYDER S H. Nitric oxide: A physiologic messenger molecule [J]. Annu Rev Biochem, 1994, 63: 175-195.

[10] HOU Y C, JANCZUK A, WANG P G. Current trends in the development of nitric oxide donors [J]. Curr Pharm Des, 1999, 5 (6): 417-441.

[11] MIRANDA K M. The chemistry of nitroxyl (HNO) and implications in biology [J]. Coordin Chem Rev, 2005, 249 (3/4): 433-455.

[12] GLADWIN M T, SCHECHTER A N, KIM-SHAPIRO D B, et al. The emerging biology of the nitrite anion [J]. Nat Chem Biol, 2005, 1 (6): 308-314.

[13] CHACHLAKI K, GARTHWAITE J, PREVOT V. The gentle art of saying NO: How nitric oxide gets things done in the hypothalamus [J]. Nat Rev Endocrinol, 2017, 13 (9): 521-535.

[14] ORTEGA MATEO A, AMAYA ALEIXANDRE DE A. Nitric oxide reactivity and mechanisms involved in its biological effects [J]. Pharmacol Res, 2000, 42 (5): 421-427.

[15] CALABRESE E J, AGATHOKLEOUS E, DHAWAN G, et al. Nitric oxide and hormesis [J]. Nitric Oxide, 2023, 133: 1-17.

[16] XU W M, LIU L Z, LOIZIDOU M, et al. The role of nitric oxide in cancer [J]. Cell Res, 2002, 12 (5/6): 311-320.

[17] BURKE A J, SULLIVAN F J, GILES F J, et al. The yin and yang of nitric oxide in cancer progression [J]. Carcinogenesis, 2013, 34 (3): 503-512.

[18] FAHEY J M, GIROTTI A W. Nitric oxide antagonism to anti-glioblastoma photodynamic therapy: mitigation by inhibitors of nitric oxide generation [J]. Cancers (Basel), 2019, 11 (2): 231.

[19] FAHEY J M, GIROTTI A W. The negative impact of cancer cell nitric oxide on photodynamic therapy [J]. Methods Mol Biol, 2022, 2451: 21-31.

[20] KIM J, FRANCIS D M, SESTITO L F, et al. Thermosensitive hydrogel releasing nitric oxide donor and anti-CTLA-4 micelles for anti-tumor immunotherapy [J]. Nat Commun, 2022, 13 (1): 1479.

[21] ZHANG X, JIN L, TIAN Z, et al. Nitric oxide inhibits autophagy and promotes apoptosis in hepatocellular carcinoma [J]. Cancer Sci, 2019, 110 (3): 1054-1063.

[22] TANG B, ZHUO L, GE J, et al. A surfactant-free route to single-crystalline CeO_2 nanowires [J]. Chemical Communications (Cambridge, England), 2005 (28): 3565-3567.

[23] SI R, FLYTZANI-STEPHANOPOULOS M. Shape and crystal-plane effects of nanoscale ceria on the activity of Au-CeO catalysts for the water-gas shift reaction [J]. Angew Chem Int Edit, 2008, 47 (15): 2884-2887.

[24] DOWDING J M, DOSANI T, KUMAR A, et al. Cerium oxide nanoparticles scavenge nitric oxide radical (NO)[J]. Chemical Communications (Cambridge, England), 2012, 48 (40): 4896-4898.

[25] HU F X, XIE J L, BAO S J, et al. Shape-controlled ceria-reduced graphene oxide nanocomposites toward high-sensitive in situ detection of nitric oxide [J]. Biosens Bioelectron, 2015, 70: 310-317.

[26] LUCHSINGER B P, RICH E N, GOW A J, et al. Routes to S-nitroso-hemoglobin formation with heme redox and preferential reactivity in the beta subunits [J]. P Natl Acad Sci USA, 2003, 100 (2): 461-466.

[27] GROSS S S, LANE P. Physiological reactions of nitric oxide and hemoglobin: A radical rethink [J]. P Natl Acad Sci USA, 1999, 96 (18): 9967-9969.

[28] WIGHTMAN R M, MAY L J, MICHAEL A C. Detection of dopamine dynamics in the brain [J]. Analytical Chemistry, 1988, 60 (13): 769A-779A.

[29] KUTNINK M A, HAWKES W C, SCHAUS E E, et al. An internal standard method for the unattended high-performance liquid chromatographic analysis of ascorbic acid in blood components [J]. Analytical Biochemistry, 1987, 166 (2): 424-430.

[30] ZHAI B, HU F, JIANG X, et al. Inhibition of Akt reverses the acquired resistance to sorafenib by switching protective autophagy to autophagic cell death in hepatocellular carcinoma [J]. Mol Cancer Ther, 2014, 13 (6): 1589-1598.

[31] CHIU J, TANG Y F, YAO T J, et al. The use of single-agent sorafenib in the treatment of advanced hepatocellular carcinoma patients with underlying child-pugh B liver cirrhosis [J]. Cancer-Am Cancer Soc, 2012, 118 (21): 5293-5301.

[32] WILHELM S M, CARTER C, TANG L, et al. BAY 43-9006 exhibits broad spectrum oral antitumor activity and targets the RAF/MEK/ERK pathway and receptor tyrosine kinases involved in tumor progression and angiogenesis [J]. Cancer Res, 2004, 64 (19): 7099-7109.

[33] CIAMPORCERO E, MILES K M, ADELAIYE R, et al. Combination strategy targeting VEGF and HGF/c-met in human renal cell carcinoma models [J]. Mol Cancer Ther, 2015, 14 (1): 101-110.

[34] SANOFF H K, CHANG Y, LUND J L, et al. Sorafenib effectiveness in advanced hepatocellular carcinoma [J]. Oncologist, 2016, 21 (9): 1113-1120.

[35] MARTINEZ DE LA ESCALERA L, KYROU I, VRBIKOVA J, et al. Impact of gut hormone FGF-19 on type-2 diabetes and mitochondrial recovery in a prospective study of obese diabetic women undergoing bariatric surgery [J]. BMC Med, 2017, 15 (1): 34.

[36] JONES S. Mini-review: Endocrine actions of fibroblast growth factor 19 [J]. Mol Pharm, 2008, 5 (1): 42-48.

[37] ZHAO H K, LV F L, LIANG G Z, et al. FGF19 promotes epithelial-mesenchymal transition in hepatocellular carcinoma cells by modulating the GSK3 beta/beta-catenin signaling cascade via FGFR4 activation [J]. Oncotarget, 2016, 7 (12): 13575-13586.

[38] KAIBORI M, SAKAI K, ISHIZAKI M, et al. Increased FGF19 copy number is frequently

detected in hepatocellular carcinoma with a complete response after sorafenib treatment [J]. Oncotarget, 2016, 7 (31): 49091-49098.

[39] GAO L, WANG X, TANG Y, et al. FGF19/FGFR4 signaling contributes to the resistance of hepatocellular carcinoma to sorafenib [J]. J Exp Clin Cancer Res, 2017, 36 (1): 8.

[40] LLOVET J M, RICCI S, MAZZAFERRO V, et al. Sorafenib in advanced hepatocellular carcinoma [J]. New Engl J Med, 2008, 359 (4): 378-390.

[41] YOSHIDA M, YAMASHITA T, OKADA H, et al. Sorafenib suppresses extrahepatic metastasis de novo in hepatocellular carcinoma through inhibition of mesenchymal cancer stem cells characterized by the expression of CD90 [J]. Sci Rep-Uk, 2017, 7 (1): 11292.

[42] VAN MALENSTEIN H, DEKERVEL J, VERSLYPE C, et al. Long-term exposure to sorafenib of liver cancer cells induces resistance with epithelial-to-mesenchymal transition, increased invasion and risk of rebound growth [J]. Cancer Lett, 2013, 329 (1): 74-83.

[43] VILLANUEVA A, LLOVET J M. Second-line therapies in hepatocellular carcinoma: Emergence of resistance to sorafenib [J]. Clin Cancer Res, 2012, 18 (7): 1824-1826.

[44] FARAZI P A, DEPINHO R A. Hepatocellular carcinoma pathogenesis: From genes to environment [J]. Nat Rev Cancer, 2006, 6 (9): 674-687.

[45] MEDEIROS R, MORAIS A, VASCONCELOS A, et al. Endothelial nitric oxide synthase gene polymorphisms and genetic susceptibility to prostate cancer [J]. Eur J Cancer Prev, 2002, 11 (4): 343-350.

[46] ROYLE J S, ROSS J A, ANSELL I, et al. Nitric oxide donating nonsteroidal anti-inflammatory drugs induce apoptosis in human prostate cancer cell systems and human prostatic stroma via caspase-3 [J]. J Urology, 2004, 172 (1): 338-344.

[47] SAWEY E T, CHANRION M, CAI C, et al. Identification of a therapeutic strategy targeting amplified FGF19 in liver cancer by Oncogenomic screening [J]. Cancer Cell, 2011, 19 (3): 347-358.

[48] DESNOYERS L R, PAI R, FERRANDO R E, et al. Targeting FGF19 inhibits tumor growth in colon cancer xenograft and FGF19 transgenic hepatocellular carcinoma models [J]. Oncogene, 2008, 27 (1): 85-97.

[49] ZHOU M, WANG X Y, PHUNG V, et al. Separating tumorigenicity from bile acid regulatory activity for endocrine hormone FGF19 [J]. Cancer Res, 2014, 74 (12): 3306-3316.

[50] DEGIROLAMO C, SABBA C, MOSCHETTA A. Therapeutic potential of the endocrine fibroblast growth factors FGF19, FGF21 and FGF23 [J]. Nature Reviews Drug Discovery, 2016, 15 (1): 51-69.

[51] ZHOU M, LEARNED R M, ROSSI S J, et al. Engineered fibroblast growth factor 19 reduces

liver injury and resolves sclerosing cholangitis in Mdr2-deficient mice [J]. Hepatology, 2016, 63 (3): 914-929.

[52] CABALLANO-INFANTES E, CAHUANA G M, BEDOYA F J, et al. The role of nitric oxide in stem cell biology [J]. Antioxidants (Basel), 2022, 11 (3): 497.

[53] KATUSIC Z S, CAPLICE N M, NATH K A. Nitric oxide synthase gene transfer as a tool to study biology of endothelial cells [J]. Arterioscl Throm Vas, 2003, 23 (11): 1990-1994.

[54] GANTNER B N, LAFOND K M, BONINI M G. Nitric oxide in cellular adaptation and disease [J]. Redox Biol, 2020, 34: 101550.

[55] ARCHER S. Measurement of nitric oxide in biological models [J]. FASEB J, 1993, 7 (2): 349-360.

[56] CISZEWSKI A, MILCZAREK G. A new nafion-free bipolymeric sensor for selective and sensitive detection of nitric oxide [J]. Electroanalysis, 1998, 10 (11): 791-793.

[57] WANG Y, LI Q, HU S. A multiwall carbon nanotubes film-modified carbon fiber ultramicroelectrode for the determination of nitric oxide radical in liver mitochondria [J]. Bioelectrochemistry, 2005, 65 (2): 135-142.

[58] PEREIRA-RODRIGUES N, ALBIN V, KOUDELKA-HEP M, et al. Nickel tetrasulfonated phthalocyanine based platinum microelectrode array for nitric oxide oxidation [J]. Electrochem Commun, 2002, 4 (11): 922-927.

[59] GU H Y, YU A M, YUAN S S, et al. Amperometric nitric oxide biosensor based on the immobilization of hemoglobin on a nanometer-sized gold colloid modified Au electrode [J]. Anal Lett, 2002, 35 (4): 647-661.

[60] BEDIOUI F, QUINTON D, GRIVEAU S, et al. Designing molecular materials and strategies for the electrochemical detection of nitric oxide, superoxide and peroxynitrite in biological systems [J]. Phys Chem Chem Phys, 2010, 12 (34): 9976-9988.

[61] GRIVEAU S, BEDIOUI F. Overview of significant examples of electrochemical sensor arrays designed for detection of nitric oxide and relevant species in a biological environment [J]. Anal Bioanal Chem, 2013, 405 (11): 3475-3488.

[62] MADAGALAM M, BARTOLI M, TAGLIAFERRO A. A short overview on graphene and graphene-related materials for electrochemical gas sensing [J]. Materials (Basel), 2024, 17 (2): 303.

[63] XU Y, SHENG K, LI C, et al. Self-assembled graphene hydrogel via a one-step hydrothermal process [J]. ACS Nano, 2010, 4 (7): 4324-4330.

[64] WALKER E, WANG J, HAMDI N, et al. Selective detection of extracellularglutamate in brain tissue using microelectrode arrays coated with over-oxidized polypyrrole [J]. Analyst, 2007,

132 (11): 1107-1111.

[65] BARBOSA R M, LOURENÇO C F, SANTOS R M, et al. Chapter twenty-in vivo real-time measurement of nitric oxide in anesthetized rat brain [J]. Methods in Enzymology, 2008, 441: 351-367.

6 CGMA/纸夹芯电化学传感器在细胞过氧化氢检测中的应用

6.1 引　言

由于纸是一种三维（3D）网状纤维的纤维素，其在生物传感器器件制造中的作用引起了人们的极大关注[1,2]。易用性和存储能力使其成为制造一次性生物传感器最有潜在经济价值的生物相容性材料。此外，多孔纤维素纸的网格可以充当毛细血管，不需要主动泵送即可吸干水溶液。因此，生物分析领域的研究人员已经投入了大量的努力来构建高灵敏度、操作简单和低成本的纸质分析器件[3,4]。研究表明纸质分析器件可用于不同的检测，如比色、化学发光、电化学和表面增强拉曼散射等[5-9]。通常情况下，基于纸质分析器件的传感器被用于蛋白质分析。例如，Zang 等人基于纸的 3D 微流控模型可以检测癌症标志物、α-胎蛋白和癌抗原的夹心免疫[10]。Nie 等人介绍了一种基于纸的葡萄糖传感器用于血糖检测，其具有低成本优势[11]。

细胞作为生命的基本功能单位，基于细胞的生化分析为生物学和医学问题提供了重要的信息。因此，建立一种稳定、简单、灵敏的细胞代谢监测传感器对于疾病的监测与预防具有重要作用。然而，目前，基于细胞的检测大多依赖于二维（2D）细胞培养系统，尚不能完整地呈现体内细胞生长的结构、功能和生理能力。微加工可以为细胞 3D 培养创造结构复杂的支架[12-14]，然而，这些方法需要生物实验室中不常见的仪器。基于此，"细胞-凝胶-纸"3D 细胞培养系统取得了突破性进展，该系统是由浸入水凝胶中的细胞混悬液的纸层组成，用于分析细胞中分子和基因作用机制[15-17]。在此研究工作中，开发一种基于纸质的分析器件并结合 3D 细胞培养平台，该平台用于在模拟体内细胞生长的环境条件且可以对细胞活性进行原位监测。为了证明细胞纸质传感器的生物学意义，我们检测了人类肿瘤细胞中 H_2O_2 的产生，因为 H_2O_2 在多种细胞信号通路中发挥重要作用，影响细胞增殖与迁移等[18,19]。由于其具有重要的生物学意义，人们已经设计和构建出生物传感器来检测过氧化氢[20]。酶，如 HRP 和过氧化氢酶，已经被用于固

定在电极上构建 H_2O_2 传感器[21,22]。随着对纳米材料技术不断深入的研究，非酶传感器材料逐渐成为酶传感器的经济性替代者[23-25]。例如，在碳纳米管（CNTs）[24]、介孔碳[26]或层状石墨烯[25]上合成了一种人工过氧化物酶普鲁士蓝，用以实现 H_2O_2 检测。由于二氧化锰（MnO_2）纳米颗粒具有优异的催化能力，使其成为另一种具有吸引力的 H_2O_2 传感器的无机氧化物材料[27]。Zhang 等人报道了基于全氟磺酸膜（nafion）和微球 MnO_2 修饰玻璃碳电极直接电催化氧化 H_2O_2 的方法[28]。然而，二氧化锰/氧化石墨烯纳米复合功能化玻璃碳电极实现了碱性介质中 H_2O_2 的检测[29]。最近报道发现 MnO_2 可以在碳纳米管衍生的石墨烯上生长构建了用于检测过氧化氢的电化学生物传感器[30,31]。然而，从碳纳米管上剥离石墨烯层应该在 $KMnO_4$ 和浓 H_2SO_4 的溶液中进行，温度为 80 ℃，时间为 60 min。从 SEM 表征可以看出，合成材料的形貌与碳纳米管相似，无法观察到纤细的簇状纳米颗粒网格。碳纳米管/石墨烯气凝胶由于其优异的导电性和高特异性表面积质荷比，已被应用于超级电容器和生物传感领域[32-34]。因此，我们期望在碳纳米管/石墨烯气凝胶上生长二氧化锰纳米颗粒，以实现检测细胞分泌的 H_2O_2 电化学活性。虽然过氧化氢电化学传感器已经有许多报道，但是大多数传感器使用玻璃碳电极和氧化铟锡（ITO）玻璃在标准的三电极系统中测量 H_2O_2[35-37]。由于将它们作为经济的一次性检测装置的候选物是不切实际的。此外，只有少数方法能够实现活细胞分泌的 H_2O_2 的原位检测，特别是选择性和定量检测[22,24,35]。由于细胞生物学在医学科学中具有重要地位，人们迫切需要一种简单的、一次性装置来实现传统三电极电化学分析系统的全部功能。如果该设备能够在 3D 细胞生长模型中检测 H_2O_2，同时提供可靠的病理诊断数据，人们的热情只会增强。在本研究中，将碳纳米管/石墨烯/二氧化锰复合功能化碳纸电极夹在蜡印纸中，制作了一个基于碳纳米管/石墨烯/二氧化锰气凝胶（CGMA）电极/细胞-纸夹芯装置，蜡印纸作为 3D 基质维持细胞生长，实现了对人体细胞释放过氧化氢的原位监测，揭示了 CGMA 电极/纸装置在细胞生物学研究和药物筛选中的潜在应用价值。

6.2 过氧化氢电化学生物传感器

6.2.1 碳纳米管/石墨烯/二氧化锰气凝胶（CGMA）的制备

在这项工作中使用的氧化石墨烯是选用 Hummers 方法将石墨进行改性制备而成[38]。本研究中使用的碳纳米管是在 HNO_3 中回流 24 h，用二次水洗涤并离心几次后，然后放置于真空烘箱中干燥。为了合成 CNT/石墨烯/MnO_2 水凝胶，采用超声法制备了 2 mg/mL 碳纳米管悬浮液，具体步骤如下：（1）取 20 mg 的 CNT 在 10 mL 去离子水中超声处理约 6 h，然后再加入 8 mL 去离子水，2 mL 氧化石墨烯（GO，10 mg/mL）和 200 mg $KMnO_4$ 晶体加入上述超声分散液中；（2）上述包含碳纳米管（1 mg/mL）、氧化石墨烯（1 mg/mL）和 $KMnO_4$（10 mg/mL）的混合物在室温下搅拌 16 h，反应混合物用去离子水洗涤并离心几次，然后收集沉淀；（3）用 5 mL 去离子水重悬浮沉淀物（2 mg/mL）放入玻璃瓶中，与 500 mL 抗坏血酸溶液（100 mg/mL）在 50 ℃下充分混合 15 h，形成 CNT/石墨烯/MnO_2 水凝胶，合成流程如图 6.1 所示；（4）所得样品用去离子水洗涤数次，然后冷冻干燥 24 h，使水分完全去除；（5）获得 CNT/石墨烯/MnO_2 气凝胶，并标记为 CGMA。此外，实验过程中制备了未经抗坏血酸溶液处理的冷冻干燥 CNTs/石墨烯/MnO_2（CGM）为后续进行功能比较。

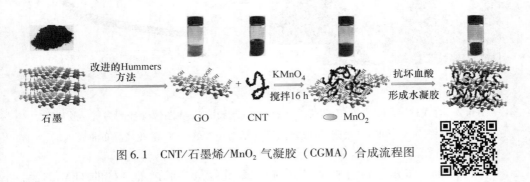

图 6.1 CNT/石墨烯/MnO_2 气凝胶（CGMA）合成流程图

图 6.1 彩图

6.2.2 碳纳米管/石墨烯/二氧化锰气凝胶（CGMA）电极的表征

为了对 CGMA 纳米复合电极的微观形貌进行观察，采用扫描电镜（SEM，JSM-6510LV，日本）获得样品的微观结构。采用 X 射线光电子能谱（XPS）、热模分析（Thermo）。氮吸附测量选用 Quanta chrome NoVa 1200e 仪器进行分析。碳的比表面积和孔径分布采用 Brunauer、Emmett 和 Teller（BET）法计算。为了表征纳米复合材料的电化学性能，将 2.5 mg CNT/石墨烯/MnO_2 气凝胶分散在 500 mL 乙醇中，并将 5 mL 悬浮液（5 mg/mL）滴加到玻碳电极（GCE，直径 3 mm，CH Instruments）的表面。在 CHI 760e 电化学工作站上，对含有 50 mmol/L $K_3Fe(CN)_6$ 的 0.5 mol/L KCl 溶液进行循环伏安和电化学阻抗谱（EIS）测量。图 6.2（a）展示了 CNT/石墨烯/MnO_2 气凝胶的 SEM 图像，具有典型的石墨烯褶皱纸状形态,该样品的局部放大图像如图 6.2(a) 中插图所示。在碳纳米管/石墨烯/二氧化锰气凝胶中发现了一些中孔和大孔,表明形成了多孔结构的纳米复合材料。通过氮气吸附和解吸实验,定量表征了碳纳米管/石墨烯/二氧化锰气凝胶的表面积和孔径分布。氮吸附和脱附等温线如图 6.2(b) 所示。利用这些等温线通过 BET 计算可得合成气凝胶的碳表面积为 133.1 m^2/g。此外,孔径分布实验显示,在 14 nm 处有一个尖峰,在 30~85 nm 处有一个宽峰,表明气凝胶富含分层孔隙(见图 6.2(b) 中插图)。

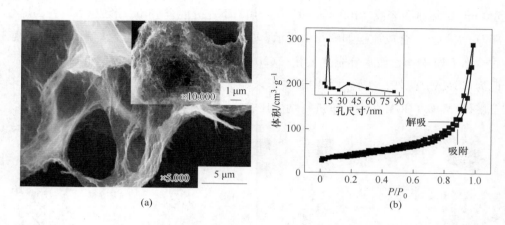

图 6.2　合成材料碳纳米管/石墨烯/二氧化锰气凝胶的 SEM 图像(插图为气凝胶局部结构放大图)(a) 和氮气吸附/解吸等温线(插图为气凝胶 N_2 的孔径分布曲线)(b)

图 6.3(a) 展示了碳纳米管/石墨烯/二氧化锰气凝胶内部结构的 TEM 图像,气凝胶的微观结构为一个平面透明薄片,上面附着有宽度为 20~30 nm 的碳纳米管。此外,在石墨烯薄片和碳纳米管侧壁上可以观察到分布均匀的黑点。由高分辨率 TEM 图像可知,均匀黑点的直径为 5 nm。此外,在一张 TEM 照片中观察到三个间距为 0.21 nm、0.24 nm 和 0.37 nm 的晶格条纹如图 6.3(b)~(d) 所示,可以分别指示为(112)、(111) 和(002) a 型 MnO_2 和石墨烯的晶体面。

碳纳米管/石墨烯/二氧化锰气凝胶的典型 XPS 光谱如图 6.4(a) 所示。从低结合能到高结合能,可以观察到 C1s, O1s 和 Mn ($2p^{3/2}$, $2p^{1/2}$) 的峰。有两个以 643.2 eV 和 654.6 eV 为中心的峰,从 Mn 的 2p3 峰产生 11.4 eV 的自旋轨道分裂(见图 6.4(b))。根据之前的报道,在 643.2 eV 和 654.6 eV 处的峰分别为 Mn $2p^{3/2}$ 和 Mn $2p^{1/2}$[30,39]。基于 SEM、TEM 和 XPS 表征得到的信息,我们得出结论,通过图 6.1 的反应流程图获得 CNT/石墨烯/MnO_2 气凝胶已经合成成功。

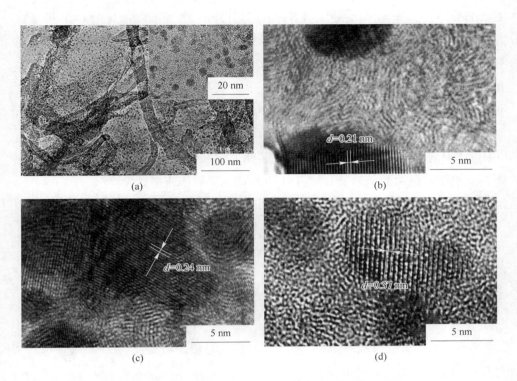

图 6.3 合成的碳纳米管/石墨烯/二氧化锰气凝胶的 TEM 图像（a）和
二氧化锰在碳纳米管/石墨烯/二氧化锰气凝胶中的晶格（b）~（d）

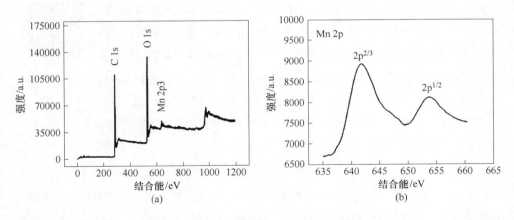

图 6.4 CNT/石墨烯/MnO_2 气凝胶的 XPS 光谱（a）和 MnO_2 的 Mn 2p XPS 光谱（b）

接下来，我们通过表征 CNT/石墨烯/MnO_2 气凝胶功能化的玻璃碳电极（GCE），研究了 CNT/石墨烯/MnO_2 气凝胶的电化学性能。如图 6.5(a) 所示，

在 [Fe(CN)$_6$]$_3$ 电解质溶液中，CNT/石墨烯/MnO$_2$ 气凝胶功能化的 GCE 中呈现出最高的电流密度，这表明 CNT/石墨烯气凝胶结构和修饰的纳米尺寸金属纳米催化剂的电化学活性表面积得到了改善。显然，抗坏血酸处理可以形成纳米大小的二氧化锰修饰的 CNT/石墨烯气凝胶，其分层孔径和表面积为 133.1 m^2/g，这可能导致电极的实际表面积扩大。图 6.5(b) 中的 EIS 曲线进一步表明，CNT/石墨烯/MnO$_2$ 气凝胶化 GCE 的电荷转移电阻为 8Ω，远小于 CNT/石墨烯/MnO$_2$（未抗坏血酸处理）/GCE(82 Ω) 和裸露 GCE 电极 (110 Ω)，从而保证了 CNT/石墨烯/MnO$_2$ 气凝胶纳米复合材料向 GCE 的电子转移速率更快。

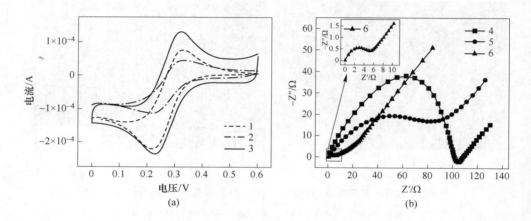

图 6.5　裸玻璃碳电极（GCE）、CNT 石墨烯/MnO$_2$（未经抗坏血酸处理）修饰的玻璃碳电极（CGM-GCE）、CNT/石墨烯/MnO$_2$ 气凝胶修饰的玻璃碳电极（CGMA-GCE 分别在含有 50 mmol/L K$_3$Fe(CN)$_6$、0.5 mol/L KCl 溶液中扫描速率为 10 mV/s 的 CV 曲线（a）和裸 GCE、GGM-GCE、CGMA-GCE 电极在含有 50 mmol/L K$_3$Fe(CN)$_6$ 的 0.5 mol/L KCl 溶液中的 EIS（b）

1—裸玻璃碳电极（GCE）；2—CNT/石墨烯/MnO$_2$（未经抗坏血酸处理）修饰的玻璃碳电极（CGM-GCE）；3—CNT/石墨烯/MnO$_2$ 气凝胶修饰的玻璃碳电极（CGMA-GCE）；4—裸 GCE 电极；5—GGM-GCE 电极；6—CGMA-GCE 电极

上述结果表明，碳纳米管/石墨烯/二氧化锰气凝胶可能会产生具有增强催化活性的电极，这可能与石墨烯片具有平面结构的高表面积有关。该结构为二氧化锰纳米颗粒的沉积提供了大量的锚定位点，而在石墨烯片上组装的碳纳米管改善了从石墨烯平面到电极的电子转移。此外，气凝胶结构有利于分层孔隙的形成，防止纳米级颗粒的聚集。

6.2.3 CGMA 电极/纸夹芯装置的制作

如图 6.6(a) 所示,在蜡模型滤纸上绘制疏水和亲水区域。滤纸的亲水性区域设计用于分析剂装载和细胞生长。用银胶作为电极导线。用切纸机将碳纸(CP)切成所需的尺寸。然后在滤纸的疏水区域用银胶连接 CP 电极(CPE)的一端,在滤纸的亲水性区域留下 2 mm 长的 CPE。因此,在所有的实验中,工作电极的尺寸为 1×2 mm^2,对电极的尺寸为 4×2 mm^2。然后,将滤纸夹在 CPE 之间以组装双电极体系的电化学设备(见图 6.6(b))。因为蜡模型滤纸可以将水溶液吸到湿电极上,所以测试只需要很低的样本量。在接下来的实验中,如果未指定,在亲水性区域均是加入 100 mL PBS 进行电化学测量。为了使 CPE 工作电极功能化,将含有 CNT/石墨烯/MnO$_2$ 气凝胶的 2 mL 悬浮液(5 mg/mL)沉积在 CPE 表面,然后滴加 1 μL 的 Nafion 乙醇稀释液(Nafion 与乙醇的比例为 1∶30, V/V)。干燥 15 min 后,CPE/纸/CPE 夹心装置即可使用(见图 6.6(c))。采用类似的方法,制备了 CNT/石墨烯/MnO$_2$(未经抗坏血酸处理)功能化 CPE 作为对照。

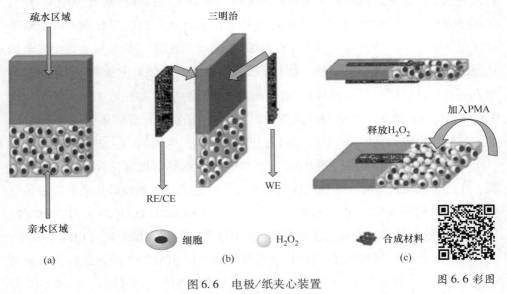

图 6.6 电极/纸夹心装置

(a) 滤纸疏水区和亲水区蜡纹;(b) 碳纸电极和蜡纹纸夹心(RE/CE 为参比电极/对电极,WE 为工作电极);(c) 组装电极/纸装置和原位 H$_2$O$_2$ 检测(PMA 为佛波醇 12-十四酸酯 13-乙酸酯)

6.2.4 CPGA 电极/纸夹芯装置的电化学表征

CPE/纸/CPE 夹芯装置电化学测量的设置如图 6.6 所示。通过调节电势从

−0.2 到 −0.7 V，优化了器件对 H_2O_2 传感器的电流响应。所有电位测量和报告均是以碳纸作为对电极进行的。在工作电位 −0.5 V 下，对 CPE/纸/CPE 夹芯装置与碳纸对电极的灵敏度进行了电流监测。此外，通过测量多巴胺（DA）、尿酸（UA）和抗坏血酸（AA）的电流反应，证明了 CPE/纸/CPE 夹芯装置的特异性。所有测量均在 pH = 7.0 的 0.1 mol/L PBS 溶液中进行测试，并至少独立重复三次。

6.3 基于 CGMA 电极/纸夹芯装置的电化学传感器检测过氧化氢

通过完全自组装碳纸电极/纸夹芯装置（CPE/纸）检测了过氧化氢的响应能力。在 0.1 mol/L、pH = 7.0 的 PBS 溶液中加入 5 mmol/L 的过氧化氢，CNT/石墨烯/MnO_2 气凝胶功能化电极的循环伏安曲线（CV）显示出随着电流的减小，峰值还原电流出现急剧增加（见图 6.7(a)）。然而未经抗坏血酸处理的 CNT/石墨烯/MnO_2 气凝胶修饰的 CPE 和裸露的 CPE 的循环伏安曲线变化则可以忽略不计（见图 6.7(b) 和 (c)），CNT/石墨烯/MnO_2 气凝胶修饰的电极表现出显著的还原过氧化氢的电极催化行为。为了证明完全自组装的装置电极对 H_2O_2 的循环伏安的特异性响应，添加了 H_2O_2 清除剂过氧化氢酶[40]。如图 6.7(d) 所示，在 0.1 mol/L、pH 值为 7.0 的 PBS 溶液中，2 mmol/L 的 H_2O_2 导致还原峰电流增加，加入过氧化氢酶后可使 CV 曲线返回到 PBS 对照组的电流值。

为了实现传感器实时检测 H_2O_2 的产生，通常选择对过氧化氢的电流进行监测。在记录电流信号时，选择工作电极上施加的电位是达到最低检测限和避免电化学干扰物质的关键。如图 6.8(a) 所示，0.2 mmol/L H_2O_2 在 CNT/石墨烯/MnO_2 气凝胶上的净稳态氧化还原反应（CGMA）功能化碳纸电极（CGMA/CPE）在 −0.2 V 至 −0.7 V 的电位范围内进行了记录。CGMA/CFP 的背景电流在测试电位范围内（−0.2 V 至 −0.4 V）都很低（见图 6.8(a)，黑线）。从 −0.5 V 开始，背景电流的幅度迅速增加。相比之下，CGMA/CPE 在 0.2 mmol/L H_2O_2 处理下 −0.4 V、−0.5 V、−0.6 V 和 −0.7 V 电位下的净还原电流分别为 0.4 mA、0.8 mA、1.2 mA 和 1.7 mA（见图 6.8(a)，红线）。为了获得灵敏的响应，同时有效地避免干扰，在后续实验中选择 −0.5 V 作为应用电位。

抗坏血酸（AA）、多巴胺（DA）和尿酸（UA）是生物样品中与过氧化氢共

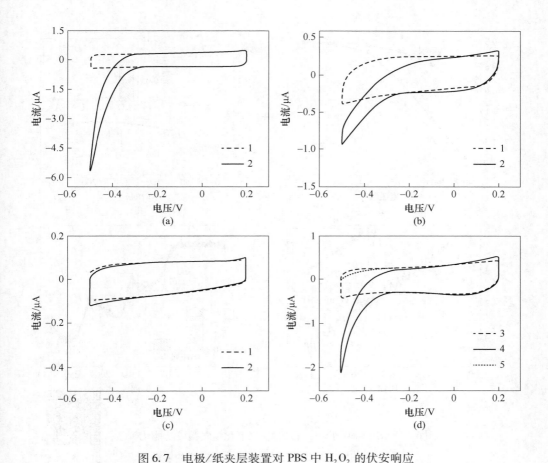

图 6.7 电极/纸夹层装置对 PBS 中 H_2O_2 的伏安响应

(a) CNT/石墨烯/MnO_2 气凝胶功能化碳纸电极；(b) CNT/石墨烯/MnO_2（未加抗坏血酸处理）

功能化碳纸电极；(c) 裸碳纸电极；(d) CNT/石墨烯/MnO_2 气凝胶-碳纸电极

1，3—不加 H_2O_2；2—加 5 mmol/L H_2O_2；4—加 2 mmol/L 的 H_2O_2；

5—在 2 mmol/L H_2O_2 中加入过氧化氢酶

存的潜在干扰物，为了证明碳纳米管/石墨烯/二氧化锰气凝胶功能化夹芯装置电极对 H_2O_2 的选择性，分别监测了该装置对 H_2O_2 和常见干扰的电流响应。如图 6.8(b) 所示，电极对 AA、DA 和 UA 的电流响应明显低于 H_2O_2，表明完全自组装的夹芯装置电极可以对产生的 H_2O_2 具有特异性响应。本研究测试的干扰浓度为 0.2 mmol/L，这是 AA、DA 和 UA 的最大生理浓度范围[41-43]，结果表明，CNT/石墨烯/MnO_2 气凝胶装置能够选择性地检测生物样品中的 H_2O_2（见图 6.8

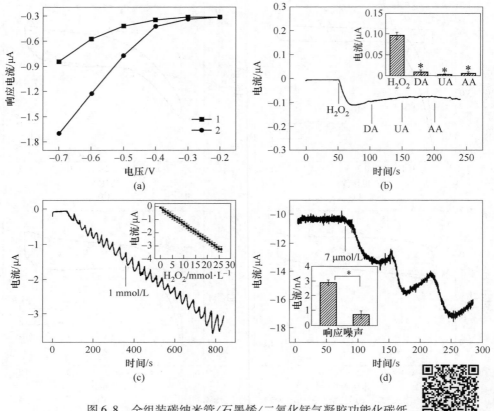

图6.8 全组装碳纳米管/石墨烯/二氧化锰气凝胶功能化碳纸电极（CGMA-CPE）/纸三明治电化学装置的表征

图6.8 彩图

(a) 以 H_2O_2 为模型，研究了电极/纸三明治装置的最佳工作电位（恒定还原电流在 PBS（pH=7.0）中测量，不含（黑色）或含（红色）0.2 mmol/L H_2O_2，电位范围为 −0.2 ~ −0.7 V）；(b) 干扰物质对生物传感器响应的影响（H_2O_2(1.0 mmol/L)，多巴胺（DA，0.2 mmol/L），尿酸（UA，0.2 mmol/L）和抗坏血酸（AA，0.2 mmol/L），直方图由三次独立实验计算获得（*表示 $p<0.05$））；(c) 测定了 CGMA-CPE/纸夹芯装置在施加电位 −0.5 V 时连续加入 1 mmol/L 的 H_2O_2 时的电流响应；(d) CGMA-CPE/纸夹芯装置对 PBS 中低浓度 H_2O_2 响应的典型电流变化（插图为 7 mmol/L H_2O_2 响应时的信噪信号和电流变化的直方图（$n=3$，*表示 $p<0.05$））

(b))。图 6.8(c) 显示了 CNT/石墨烯/MnO_2 气凝胶功能化夹芯装置在 −0.5 V 的 PBS 中添加 H_2O_2 的典型安培响应。实验发现完全自组装的夹心装置电极对 H_2O_2 浓度的变化反应迅速。在扩展测量过程中，可以观察到电流下降的趋势。这可能是由于附近区域 H_2O_2 浓度降低所致。由三次独立测量计算得出的电流-剂量响应曲线如图 6.8(c) 中插图所示,该电极具有宽线性检测范围,可以达到 25 mmol/L。基于电流-剂量响应曲线斜率与电极表面积的比值,完全自组装的电极的灵敏度为 6.25 mA/(mmol·L^{-1}·cm^2)。此外,根据 3 倍的噪声比测试进行计算,该器件实现了 6.7 μmol/L 的低检测限 (见图 6.8(c))。综上可知,完全自组装的电极/纸三明治装置显示了其作为一次性装置在广泛工作范围内进行 H_2O_2 传感的潜力。电极/纸三明治结构可以检测低至 40 μL 的样品,从而显著减少了用于检测的样本量。这对于罕见的生物样本,如临床活检分析具有重要意义。

6.4 基于 CNT/石墨烯/MnO$_2$ 纳米复合物与纸夹芯装置的电化学传感器原位监测活细胞释放 H_2O_2

为了实现对细胞分泌 H_2O_2 的原位电化学检测,我们将人喉癌细胞 HEp2 直接生长在夹在碳纸电极之间基质胶 1 的浸渍纸上。人喉癌细胞 HEp2 维持在含有 10% FBS+100 U/mL 青霉素和 100 U/mL 链霉素的 RMPI 中。细胞在含 5% CO_2 的 37 ℃ 细胞培养室中孵育。为了提高细胞在滤纸上的黏附和生长,将细胞悬浮于无生长因子基质中,基质的最低浓度为 4.5×10^6 个细胞/mL。基质在 4 ℃ 时为液体,在 10 ℃ 以上迅速凝胶化。除非另有说明,否则用 1∶5 (V/V) 的冷冻细胞悬浮液稀释基质。每个基质悬浮液取 40 μL 该冷冻细胞,用移液管将其滴到预冷碳纸上。一旦基质固化,夹层双电极装置就可以进行现场电化学测量。佛波醇 12-十四酸酯 13-乙酸酯 (phorbol 12-myristate-13-acetate,PMA) 被用作触发细胞产生和释放过氧化氢的模型药物。通过药物 PMA 对细胞进行刺激,然后对释放的 H_2O_2 进行了原位监测。PMA 是一种已知能诱导人体细胞产生过氧化氢的化学物质,将 PMA 添加到含有 2×10^5 个细胞的纸凝胶上,记录了施加电位 -0.5 V 时的安培电流响应。此外,DADH-泛醌氧化还原酶抑制剂,二苯基氯化碘盐 (DPI)[44]、H_2O_2 清除剂过氧化氢酶[40] 和 PMA 被用于监测细胞原位产生的 H_2O_2 水平。如图 6.9(a) 所示,5 mg/mL 的 PMA 刺激可以显著提高还原峰电流 (线:细胞响应 4),然而在对照组和二甲基亚砜 (DMSO) 溶剂对照组中没有电流响应 (线:控制;线:溶剂对照组)。由于 DPI 可以抑制线粒体呼吸产生 H_2O_2,因此 DPI 可以减少由 PMA 引起的电流显著变化 (线:细胞响应 2)。此外,添加过氧化氢酶也可以将过氧化氢分解成水和氧气,使 PMA 引起的峰电流增加急剧减少 (线:细胞响应 3)。

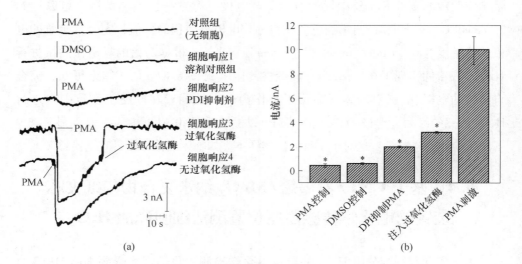

图 6.9 PMA 处理下 CNT/石墨烯/MnO₂ 气凝胶修饰的碳纸电极（CGMA-CPE）/无细胞的纸夹芯装置（对照组：无细胞）、DMSO 加入后培养的细胞 CGMA-CPE/纸夹芯装置（细胞响应 1：溶剂对照组）、PMA 和 DPI 共同处理细胞后的 CGMA-CPE/纸夹芯装置的电流响应（细胞响应 2：PDI 抑制剂）、PMA 与过氧化氢酶共同处理细胞后 CMA-CPE/纸夹芯装置电流响应（细胞响应 3：过氧化氢酶）和 CMA-CPE/纸夹芯装置与 PMA 注射培养细胞（细胞响应 4：无过氧化氢酶）(a) 与对 (a) 图中每组实验进行三次独立实验后测得的电流变化直方图 (b)($n=3$，* 表示 $p<0.05$)

图 6.9(b) 为 3 次独立测试得出的电流变化直方图。4 mL PMA（5 mg/mL）可以刺激细胞分泌 H_2O_2，其特征是电流变化为 9.94 nA。每个细胞释放的胞外 H_2O_2 分子数（No）可根据 Guo 等人的公式计算[25]：

$$No = [\Delta R/(k \times A) \times V] \times N_A/N_c$$

式中，ΔR 为电流响应；k 为传感器平台灵敏度；A 为电极表面积；V 为电解液体积；N_A 为阿伏伽德罗常数（6.02×10^{23}/mol）；N_c 为细胞数量。

已知电流响应为 9.94 nA，灵敏度为 6.25 mA/(mmol·L⁻¹·cm²)，一个传感器电极面积为 2 mm²，大约 2×10^5 个细胞，电解质的体积为 40 μL 时，可以计算出 No 约为 1.06×10^{11} 个，这与文献报道的数据非常吻合[25]。

值得注意的是，电流响应表明加入 DPI 后可以将 PMA 诱导的电流变化降低到 2 nA，显示了该传感器实时监测药物作用的能力。综上所述，实验结果证实了电流响应是由喉癌 HEp2 细胞分泌的 H_2O_2 直接产生的，证明了该传感器可以实现对纸凝胶基质中生长的细胞释放的小分子进行原位监测。这种电极/纸三明治装置可以作为实时监测细胞在模拟体内条件的微环境中分泌电活性物质的有效候选设备。综上所述，我们展示了一种新型的独立式电极/纸三明治装置，通过在面对面排列的工作电极和对电极之间布置一层纸。由于该装置中的碳纸可以吸收含水样品以湿润电极进行电化学测量，因此仅需要少量样品即可进行分析。采用完全自组装的电极/纸夹芯装置对过氧化氢进行监测，该独立装置显示线性范围达 25 mmol/L，灵敏度为 6.25 mA/(mmol·L^{-1}·cm^2)，在 PBS 中 H_2O_2 的检测限为 6.7 mmol/L。由于碳纸与电极的紧密接触，可以实时监测在纸基质中生长的细胞释放的 H_2O_2。我们设想，我们设计的柔性电极/纸夹芯装置的模块化方法可以为设计用于细胞生物学研究提供低成本一次性微型生物传感器提供新的见解。

参 考 文 献

[1] LISOWSKI P, ZARZYCKI P K. Microfluidic paper-based analytical devices (μPADs) and micro total analysis systems (μTAS): Development, applications and future trends [J]. Chromatographia, 2013, 76 (19): 1201-1214.

[2] NILGHAZ A, WICAKSONO D H, GUSTIONO D, et al. Flexible microfluidic cloth-based analytical devices using a low-cost wax patterning technique [J]. Lab on a chip, 2012, 12 (1): 209-218.

[3] MARTINEZ A W, PHILLIPS S T, BUTTE M J, et al. Patterned paper as a platform for inexpensive, low-volume, portable bioassays [J]. Angew Chem Int Ed Engl, 2007, 46 (8): 1318-1320.

[4] MARTINEZ A W, PHILLIPS S T, WHITESIDES G M, et al. Diagnostics for the developing world: Microfluidic paper-based analytical devices [J]. Anal Chem, 2010, 82 (1): 3-10.

[5] ELLERBEE A K, PHILLIPS S T, SIEGEL A C, et al. Quantifying colorimetric assays in paper-based microfluidic devices by measuring the transmission of light through paper [J]. Anal Chem, 2009, 81 (20): 8447-8452.

[6] TAN S N, GE L, TAN H Y, et al. Paper-based enzyme immobilization for flow injection electrochemical biosensor integrated with reagent-loaded cartridge toward portable modular device [J]. Anal Chem, 2012, 84 (22): 10071-10076.

[7] WANG P, GE L, YAN M, et al. Paper-based three-dimensional electrochemical immunodevice based on multi-walled carbon nanotubes functionalized paper for sensitive point-of-care testing [J]. Biosens Bioelectron, 2012, 32 (1): 238-243.

[8] LIU W, CASSANO C L, XU X, et al. Laminated paper-based analytical devices (LPAD) with origami-enabled chemiluminescence immunoassay for cotinine detection in mouse serum [J]. Anal Chem, 2013, 85 (21): 10270-10276.

[9] LI B, ZHANG W, CHEN L, et al. A fast and low-cost spray method for prototyping and depositing surface-enhanced Raman scattering arrays on microfluidic paper based device [J]. Electrophoresis, 2013, 34 (15): 2162-2168.

[10] ZANG D, GE L, YAN M, et al. Electrochemical immunoassay on a 3D microfluidic paper-based device [J]. Chemical Communications (Cambridge, England), 2012, 48 (39): 4683-4685.

[11] NIE Z, DEISS F, LIU X, et al. Integration of paper-based microfluidic devices with commercial electrochemical readers [J]. Lab on a chip, 2010, 10 (22): 3163-3169.

[12] ZIÓŁKOWSKA K, KWAPISZEWSKI R, BRZÓZKA Z. Microfluidic devices as tools for mimicking the in vivo environment [J]. New Journal of Chemistry, 2011, 35 (5): 979-990.

[13] HUANG C P, LU J, SEON H, et al. Engineering microscale cellular niches for three-dimensional multicellular co-cultures [J]. Lab on a chip, 2009, 9 (12): 1740-1748.

[14] YU L, NG S R, XU Y, et al. Advances of lab-on-a-chip in isolation, detection and post-processing of circulating tumour cells [J]. Lab on a chip, 2013, 13 (16): 3163-3182.

[15] DERDA R, TANG S K, LAROMAINE A, et al. Multizone paper platform for 3D cell cultures [J]. Plos One, 2011, 6 (5): e18940.

[16] DEISS F, MAZZEO A, HONG E, et al. Platform for high-throughput testing of the effect of soluble compounds on 3D cell cultures [J]. Anal Chem, 2013, 85 (17): 8085-8094.

[17] DERDA R, LAROMAINE A, MAMMOTO A, et al. Paper-supported 3D cell culture for tissue-based bioassays [J]. Proc Natl Acad Sci U S A, 2009, 106 (44): 18457-18462.

[18] HE L, ZHU C, JIA J, et al. ADSC-Exos containing MALAT1 promotes wound healing by targeting miR-124 through activating Wnt/β-catenin pathway [J]. Biosci Rep, 2020, 40 (5): BSR20192549.

[19] MA L, ZHU W Z, LIU T T, et al. H_2O_2 inhibits proliferation and mediates suppression of migration via DLC1/RhoA signaling in cancer cells [J]. Asian Pacific journal of cancer prevention: APJCP, 2015, 16 (4): 1637-1642.

[20] CIRCU M L, AW T Y. Reactive oxygen species, cellular redox systems, and apoptosis [J]. Free Radic Biol Med, 2010, 48 (6): 749-762.

[21] VARMA S, MITRA C K. Bioelectrochemical studies on catalase modified glassy carbon paste electrodes [J]. Electrochem Commun, 2002, 4 (2): 151-157.

[22] MATHARU Z, ENOMOTO J, REVZIN A. Miniature enzyme-based electrodes for detection of

hydrogen peroxide release from alcohol-injured hepatocytes [J]. Anal Chem, 2013, 85 (2): 932-939.

[23] CHU Z, SHI L, LIU Y, et al. In-situ growth of micro-cubic prussian blue-TiO_2 composite film as a highly sensitive H_2O_2 sensor by aerosol co-deposition approach [J]. Biosens Bioelectron, 2013, 47: 329-334.

[24] ZOU Y, SUN L X, XU F. Biosensor based on polyaniline-prussian blue/multi-walled carbon nanotubes hybrid composites [J]. Biosensors and Bioelectronics, 2007, 22 (11): 2669-2674.

[25] GUO C X, ZHENG X T, LU Z S, et al. Biointerface by cell growth on layered graphene-artificial peroxidase-protein nanostructure for in situ quantitative molecular detection [J]. Adv Mater, 2010, 22 (45): 5164-5167.

[26] BAI J, QI B, NDAMANISHA J C, et al. Ordered mesoporous carbon-supported prussian blue: Characterization and electrocatalytic properties [J]. Microporous and Mesoporous Materials, 2009, 119 (1/2/3): 193-199.

[27] LIN Y, CUI X, LI L. Low-potential amperometric determination of hydrogen peroxide with a carbon paste electrode modified with nanostructured cryptomelane-type manganese oxides [J]. Electrochem Commun, 2005, 7 (2): 166-172.

[28] ZHANG L, FANG Z, NI Y, et al. Direct electrocatalytic oxidation of hydrogen peroxide based on nafion and microspheres MnO_2 modified glass carbon electrode [J]. Int J Electrochem Sc, 2009, 4 (3): 407-413.

[29] LI L, DU Z, LIU S, et al. A novel nonenzymatic hydrogen peroxide sensor based on MnO_2/graphene oxide nanocomposite [J]. Talanta, 2010, 82 (5): 1637-1641.

[30] YE D, LI H, LIANG G, et al. A three-dimensional hybrid of MnO_2/graphene/carbon nanotubes based sensor for determination of hydrogen-peroxide in milk [J]. Electrochim Acta, 2013, 109: 195-200.

[31] CHEN Y, ZHANG Y, GENG D, et al. One-pot synthesis of MnO_2/graphene/carbon nanotube hybrid by chemical method [J]. Carbon, 2011, 49 (13): 4434-4442.

[32] WANG Y, LI Z, WANG J, et al. Graphene and graphene oxide: Biofunctionalization and applications in biotechnology [J]. Trends in biotechnology, 2011, 29 (5): 205-212.

[33] LIU M, LIU R, CHEN W. Graphene wrapped Cu_2O nanocubes: Non-enzymatic electrochemical sensors for the detection of glucose and hydrogen peroxide with enhanced stability [J]. Biosensors and Bioelectronics, 2013, 45: 206-212.

[34] CHEN D, TANG L, LI J. Graphene-based materials in electrochemistry [J]. Chemical Society Reviews, 2010, 39 (8): 3157-3180.

[35] LI C, ZHANG H, WU P, et al. Electrochemical detection of extracellular hydrogen peroxide released from RAW 264.7 murine macrophage cells based on horseradish peroxidase-hydroxyapatite nanohybrids [J]. Analyst, 2011, 136 (6): 1116-1123.

[36] YAO S, XU J, WANG Y, et al. A highly sensitive hydrogen peroxide amperometric sensor based on MnO_2 nanoparticles and dihexadecyl hydrogen phosphate composite film [J]. Anal Chim Acta, 2006, 557 (1/2): 78-84.

[37] ZHAI J, ZHAI Y, WEN D, et al. Prussian blue/multiwalled carbon nanotube hybrids: Synthesis, assembly and electrochemical behavior [J]. Electroanalysis: An International Journal Devoted to Fundamental and Practical Aspects of Electroanalysis, 2009, 21 (20): 2207-2212.

[38] ZHANG J, XIONG Z, ZHAO X S. Graphene-metal-oxide composites for the degradation of dyes under visible light irradiation [J]. Journal of Materials Chemistry, 2011, 21 (11): 3634-3640.

[39] YAN J, FAN Z, WEI T, et al. Fast and reversible surface redox reaction of graphene-MnO_2 composites as supercapacitor electrodes [J]. Carbon, 2010, 48 (13): 3825-3833.

[40] KIRKMAN H N, GAETANI G F. Catalase: A tetrameric enzyme with four tightly bound molecules of NADPH [J]. Proceedings of the national academy of sciences, 1984, 81 (14): 4343-4347.

[41] DE OLIVEIRA E P, BURINI R C. High plasma uric acid concentration: Causes and consequences [J]. Diabetology & metabolic syndrome, 2012, 4: 1-7.

[42] MO J W, OGOREVC B. Simultaneous measurement of dopamine and ascorbate at their physiological levels using voltammetric microprobe based on overoxidized poly (1,2-phenylenediamine)-coated carbon fiber [J]. Analytical Chemistry, 2001, 73 (6): 1196-1202.

[43] CAPASSO G, JAEGER P, ROBERTSON W G, et al. Uric acid and the kidney: Urate transport, stone disease and progressive renal failure [J]. Curr Pharm Design, 2005, 11 (32): 4153.

[44] GALLIE D R, LE H, CALDWELL C, et al. Analysis of translation elongation factors from wheat during development and following heat shock [J]. Biochem Bioph Res Co, 1998, 245 (2): 295-300.

7 电化学生物传感器的应用展望

7.1 引 言

随着社会科学技术的高速发展,电化学生物传感技术已成为关注的热点之一。电化学作为一项强大的技术,能够实时原位检测具有高空间和时间分辨率的生物标志物[1,2],该方法可用于研究细胞、组织和器官水平的生理机制以及阐明其他复杂系统的健康和疾病状态[3]。这一领域的最新发展已经证明电化学传感器在临床和生物环境中的潜力。电化学生物传感器作为生物技术的重要分支,已成为肿瘤检测领域的一种新兴技术[4,5],其有效地解决了传统肿瘤检测技术面临的各种局限与挑战,为肿瘤的早期诊断提供了强有力的技术保障。

生物学中一个相对较新的研究领域是氧化应激的研究。在抗氧化/亚硝化应激的抗氧化防御机制研究中,这一主题越来越受到关注[6]。众所周知,虽然氧气是生物体必不可少的组成部分,但涉及氧气的生化反应可能导致活性氧代谢物的形成,从而挑战抗氧化防御系统[7]。这些活性物质的存在导致一系列反应产生自由基或活性非自由基物质,分为两类,称为活性氧(ROS)和活性氮(RNS)[8]。近年来的研究强调了 ROS/RNS 在生化过程中的重要作用。然而,由于这些物质的高反应性和短寿命,对它们的研究是困难的。电化学生物传感器作为具有高灵敏度与快响应的优势用于检测细胞中 ROS/RNS 具有一定的适用性。与单细胞研究(如贴片夹紧)或体内研究的必要参数和条件相比,在细胞培养中进行测量是验证传感器选择性与便捷性的较好方法[9]。不同的生物过程可以很容易地在活细胞中被调节或触发,例如氧化反应的产生,然后可以使用电化学传感器进行研究。一些研究报告了在细胞培养中电化学 ROS/RNS 的测量,研究最广泛的是 $O_2^{·-}$、NO 和 H_2O_2[10-12]。由于裸碳或铂基电极的灵敏度和选择性较差,大多数传感器设计涉及用导电或催化层以及聚合物材料进行表面改性,以增加表面积,增强传感器信号,并最大限度地减少干扰效应。

7.2 新型电化学生物传感器

电化学生物传感器是一种能够将生物信号转化为可检测电化学信号的装置，常常与微电子技术、信息处理技术和计算机技术等紧密结合[13]。生物传感器通过对目标物质选择性识别和灵敏检测，实现对特定物质的快速、精准定量分析。新型的电化学活性氧传感器主要集中微型化、可穿戴、可视化与智能化等方面[14]。电化学传感平台的发展推动了新型生物传感器的开发，从而实现了无标记、无损检测整个细胞的活力、功能和遗传特征。大量研究试图提高电化学传感器的灵敏度和选择性等关键的性能参数。纳米结构包括纳米颗粒、纳米管和纳米线、纳米孔、自粘单层和纳米复合材料修饰电极，以提高传感器结构的性能和效率[15]。同时利用纳米结构和电化学技术的优势，导致出现了具有高灵敏度和高分解力的传感器，具有较好的应用价值。

研究开发了一种基于电化学微流控技术和组织工程的联合使用开发肿瘤芯片（TOC），该技术可以对肿瘤组织的动态微环境、生长和细胞间相互作用进行动态监测，为肿瘤学研究提供了一个有前景的平台[16]。由于电化学生物传感器的高灵敏度和选择性、低成本、便携性、易于制造和低样品消耗，寻找强大的 TOC 分析工具促进了集成化电化学生物传感器的发展。最近的报告表明，集成电化学生物传感器在 TOC 的细胞和分子水平的癌症研究中具有广阔的应用前景。全细胞电化学传感器有望实现癌代谢物的高效、便捷、低成本检测，对癌症的早期诊治具有重要意义。最近，Suhito 等人提出了一种生物多功能平台，可以同时进行 3D 多细胞肿瘤球体形成和实时评估抗癌药物治疗[17]。采用电沉积法对氧化铟锡（ITO）玻璃电极进行了 $HAuCl_4$ 修饰。该平台由高导电性金纳米结构（HCGN）组成，可以自发形成球体，并使用电化学方法检测细胞肿瘤球的生存能力。金纳米结构的表面粗糙度降低了细胞黏附能力，从而支持自动球体形成。此外，金纳米颗粒具有高导电性、长期稳定性和高生物相容性，有利于其在电化学检测中的应用。该研究有助于揭示传感器中的电子传递机制，研究所获得的全细胞生物电化学传感器构建新理论与癌代谢物生物检测新方法，将为生物传感器的设计提供新理论，为微生物电化学系统的应用提供新思路和新方法。据报道，一种基于 DNA 四面体纳米机器模块化的生物传感器，可用于细胞内生物活性小分子超敏检测[18]。该生物传感器由三个模块组建而成，包括识别靶标的适配体模块、信号放大和输出的熵驱动模块以及运输进入细胞的 DNA 四面体模块。研究人员以三磷酸腺苷（ATP）为靶标，一旦靶标 ATP 与适配体模块结合，便可从适配体模块释放引发剂从而激活熵驱动模块，最终实现 ATP 响应信号的输出和进一步的信号扩增。这种新型纳米机器在 1 pmol/L ~ 10 nmol/L 的浓度范围内对 ATP 表现出线性反应，并有着超高灵敏度，其检测限为 0.40 pmol/L。随着电化学与生物学研究的不断深入，必将涌现出各种用于分析细胞生物学功能的电化学传感器，这将对探索生命科学的奥秘提供有力的技术支撑与帮助。

7.3 未来电化学生物传感器的发展趋势

生物传感器已广泛应用于医疗、环境监测、食品安全和工业应用等领域。在医疗方面，生物传感器可以用于快速检测疾病标记物，帮助医生进行诊断和治疗。在环境监测方面，生物传感器可以高效、精准地监测大气污染、水污染等重要指标。在食品安全方面，生物传感器可以检测食品中的有毒物质和致病菌，确保食品的品质和安全。在工业应用领域，生物传感器可以用于监测工业废水和废气处理的效果等。

（1）微型化与无线化：未来生物传感器的重要发展方向是微型化和无线化。现代生物传感器已经越来越小，但是为了更好地适应医疗、环境和食品等领域的应用需求，未来的生物传感器将更为微型化，以便更容易植入人体、材料和组织中。同时，将生物传感器无线化，可实现长时间压力监测、远程监控、自适应打印和防伪溯源等功能。

（2）联合多种技术：未来生物传感器的另一重要发展方向是联合多种技术。这些技术中包括微流控技术、生物芯片技术、纳米技术、量子点技术等。例如，纳米技术可以通过改变纳米颗粒的大小和形状来改变生物传感器的灵敏度和选择性。

（3）智能化和互联网化：未来的生物传感器将体现智能化和互联网化。智能生物传感器可利用先进算法和人工智能技术快速识别、处理和传输数据结果。生物传感器的互联网化可将从各传感器的数据集成，方便地将大量数据进行监控和导航分析。

因此，开发能够准确快速地感知活细胞的电化学平台的方法在诊断和药物发现研究中具有重要作用。目前研究已经证实，通过电化学装置监测细胞代谢物，包括检测癌细胞 RONS、细胞状态和监测干细胞的多能性和分化状态。在未来，生物传感技术可以为精确医学诊断提供有用的细胞友好分析技术。电化学传感器技术是生物学研究领域的一项新进展。

综上所述，各种物质的检测是可行的，从生物小分子（例如，RONS、DNA、蛋白质、酶和激素）到细胞水平。此外，它在灵敏度、选择性和加工时间等方面具有优势，在未来的工业中是有益的。因此，快速、非破坏性和适用的电化学传感器可以被用于大规模疾病诊断系统中，对于疾病的早期诊断与预防具有重要意义。

7.4　电化学生物传感器面临的挑战

本书综述了用于检测细胞中 RONS 的电化学生物传感器。重点介绍了纳米材料在体外传感器中的应用进展，描述了柔性传感接口的优越性。综上所述，尽管众多研究团队不断努力，但是 RONS 电化学检测和定量仍存在许多有待克服的挑战。第一，也是最重要的问题是提高传感器的灵敏度、生物相容性、抗干扰性、防污性和耐久性。电化学虽然有利于在低浓度和复杂环境下量化目标物质，但对于长期检测 ROS/RNS 来量化疾病的发生和发展，是一种可行的策略。可以使用许多策略来解决这个问题，例如：引入高催化、光催化可再生和生物相容性材料，如纳米酶、金属卟啉、纳米结构金属氧化物、自清洁和仿生材料。同时，利用数学模型对不同传感器组件之间的连接进行优化。第二，挑战来自设计和构建电极界面上的活细胞培养或构建活细胞和传感材料的共培养模型。3D 培养模型保留了在 3D 微环境中培养的细胞的特征，并允许快速和直接捕获目标分子。例如，具有优异导电性和生物相容性的 3D 导电支架，可用于长期细胞培养和实时电化学监测。水凝胶是一种透明的离子导体，具有优异的生物相容性和可忽略不计的细胞毒性，可用于构建 3D 模型，允许与传感材料共同培养活细胞。具有合理生物相容性的可拉伸和柔性传感接口，有望用于构建集细胞培养和电流传感于一体的传感器。此外，这些传感器有望实时监测机械传导引起的机械诱导生化信号。第三，迫切需要改进制备技术，以生产简单、小型化的传感器，并优化植入方法，以尽量减少植入过程中对受试者（动物和患者）的组织损伤。理想的电极应该体积小，同时保持高灵敏度。要实现长期植入而不损害受试者的健康是不容易的。第四，仿生工程，即设计具有一定生物学特征的生物传感器，例如模仿组织的力学特性，为构建临床应用的电化学传感器提供了前景。因此，体内电极与组织和细胞的力学性能相匹配，增强了稳定和持久的生物界面。这样就可以对体内的化学物质进行长期、实时和原位监测，有利于安全、准确地传输信号。考虑到生物组织（如皮肤、血管、肌肉等）的异质性，因此，模拟人体组织以实现多尺度机械匹配非常重要。

最终需要开发一种将多个电极组装在一起的集成装置，实现对多个靶点的同时检测，从而为疾病的诊断和治疗提供更准确的参考。这可以与信号传输系统相结合，该系统可以通过无线数据传输将量化信息发送到终端，从而可以立即分析实时数据，例如使用智能应用程序或现有的数据分析手段。例如，工作人员可以将一些部件集成到一个灵活的可拉伸聚合物贴片上，如多个传感纤维、调理、信号转导、处理和无线传输路径，最终允许数据无线传输到支持蓝牙的移动手持设备上。传感器集成与信号传输系统相结合的概念有助于解释长期测量中各种 ROS/RNS 水平的变化与疾病发生发展的关系。本书旨在通过全面解释当前的研究工作，对现有 ROS/RNS 检测的进展和局限性进行全面的综述，希望引起人们对 ROS/RNS 电化学技术面临的挑战的关注。我们特别关注体外和体内 ROS/RNS 电化学生物传感器的研究现状。具体而言，本书综述了目前纳米技术、仿生工程和 3D 培养技术传感器的发展，为今后的工作提供指导。

7.5 小　　结

研究 ROS/RNS 的最终目的是阐明 ROS/RNS 水平异常与临床疾病发生发展的关系。到目前为止，电化学的应用已经在多个方面得到了多年的发展，一些电化学传感器已经商业化（例如，葡萄糖传感器）。因此，ROS/RNS 传感器具有成本低、捕获快、响应快、实时分析能力强等特点，是其快速发展的重要因素。许多工作不仅为 ROS/RNS 传感器的发展提供了信息，而且展示了电化学方法作为治疗临床疾病手段的潜力。尽管取得了巨大的进步，但为了将电化学技术商业化并应用于临床，还需要解决一些挑战。为了实现商业化和临床应用，需要认真考虑准确性、安全性、便捷性、可负担性、高通量分析、快速分析、重现性和工业化生产等因素，并制定解决这些方面局限性的策略。在传感器设计中，可以使用地球上含量高、理论上催化性能高的传感材料。其次，改进制备工艺，采用标准化、环保、可控的制造工艺，实现工业化生产，保证传感器之间的可重复性。此外,传感器可以提高检测速度和精度,延长使用寿命,数据传输系统也应该改进,以维护用户数据的安全。电化学生物传感器作为高新技术的代表,其在药物研发、疾病预防和治疗等领域拥有广泛的应用前景。未来,电化学生物传感器将更好地结合微电子技术、纳米技术、量子技术等高端技术,不断实现技术升级和发展,为人类带来更多福利。

参 考 文 献

[1] MAHMUDUNNABI R G, FARHANA F Z, KASHANINEJAD N, et al. Nanozyme-based electrochemical biosensors for disease biomarker detection [J]. Analyst, 2020, 145 (13): 4398-4420.

[2] XU T L, SONG Y C, GAO W, et al. Superwettable electrochemical biosensor toward detection of cancer biomarkers [J]. Acs Sensors, 2018, 3 (1): 72-78.

[3] YUNUS G, SINGH R, RAVEENDRAN S, et al. Electrochemical biosensors in healthcare services: Bibliometric analysis and recent developments [J]. Peerj, 2023, 27 (11): e15566.

[4] TIAN X H, FENG Y H, YUAN L, et al. A dynamic electrochemical cell sensor for selective capture, rapid detection and noninvasive release of tumor cells [J]. Sens Actuators B-Chem, 2021, 330: 129345.

[5] DEZHAKAM E, KHALILZADEH B, MAHDIPOUR M, et al. Electrochemical biosensors in exosome analysis; a short journey to the present and future trends in early-stage evaluation of cancers [J]. Biosens Bioelectron, 2023, 222: 114980.

[6] FINKEL T, HOLBROOK N J. Oxidants, oxidative stress and the biology of ageing [J]. Nature, 2000, 408 (6809): 239-247.

[7] BUETLER T M, KRAUSKOPF A, RUEGG U T. Role of superoxide as a signaling molecule [J]. Physiology, 2004, 19 (3): 120-123.

[8] BAUER G. Reactive oxygen and nitrogen species: Efficient, selective, and interactive signals during intercellular induction of apoptosis [J]. Anticancer Res, 2000, 20 (6 B): 4115-4139.

[9] DUMITRESCU E, ANDREESCU S. Bioapplications of electrochemical sensors and biosensors [J]. Method Enzymol, 2017, 589: 301-350.

[10] YUAN L, LIU S, TU W, et al. Biomimetic superoxide dismutase stabilized by photopolymerization for superoxide anions biosensing and cell monitoring [J]. Analytical Chemistry, 2014, 86 (10): 4783-4790.

[11] KIM M Y, NAVEEN M H, GURUDATT N G, et al. Detection of nitric oxide from living cells using polymeric zinc organic framework-derived zinc oxide composite with conducting polymer

[J]. Small (Weinheim an der Bergstrasse, Germany), 2017, 13 (26): 1700502.

[12] DONG W, REN Y, BAI Z, et al. Fabrication of hexahedral Au-Pd/graphene nanocomposites biosensor and its application in cancer cell H_2O_2 detection [J]. Bioelectrochemistry, 2019, 128: 274-282.

[13] ZHANG Y, ZHOU N. Electrochemical biosensors based on micro-fabricated devices for point-of-care testing: A review [J]. Electroanalysis, 2022, 34 (2): 168-183.

[14] NEVES M M P D, GONZáLEZ-GARCíA M B, HERNáNDEZ-SANTOS D, et al. Future trends in the market for electrochemical biosensing [J]. Curr Opin Electroche, 2018, 10: 107-111.

[15] HUANG X P, ZHU Y F, KIANFAR E. Nano biosensors: Properties, applications and electrochemical techniques [J]. J Mater Res Technol, 2021, 12: 1649-1672.

[16] LEI L J, MA B, XU C T, et al. Emerging tumor-on-chips with electrochemical biosensors [J]. TrAC-Trends in Analytical Chemistry, 2022, 153: 116640.

[17] SUHITO I R, ANGELINE N, LEE K H, et al. A spheroid-forming hybrid gold nanostructure platform that electrochemically detects anticancer effects of curcumin in a multicellular brain cancer model [J]. Small (Weinheim an der Bergstrasse, Germany), 2021, 17 (15): e2002436.

[18] YANG S, ZHAO Z, WANG B, et al. Modular engineering of a DNA tetrahedron-based nanomachine for ultrasensitive detection of intracellular bioactive small molecules [J]. ACS Appl Mater Inter, 2023, 15 (19): 23662-23670.